THE NATIONAL PARKS
OF SOUTH AFRICA

AUTHOR'S DEDICATION

To my mother Paddy Radunski

PHOTOGRAPHER'S DEDICATION

"For Barbara,
who stayed home for our children but
travelled with me in spirit."

THE NATIONAL PARKS OF SOUTH AFRICA

WHEN MAN HAS DESTROYED NATURE IT WILL BE HIS TURN TO GO;
THE BARREN EARTH WILL SWALLOW HIM UP.
ALOYSIUS HORN

PHOTOGRAPHS BY ANTHONY BANNISTER
TEXT BY RENÉ GORDON
GENERAL CONSULTANT W. ROY SIEGFRIED

NH
NEW
HOLLAND

CONSULTANTS

Professor George Branch contributed the excellent essay on Langebaan National Park. He is a professor in the Department of Zoology at the University of Cape Town, and one of the country's leading ecologists working on rocky shores and estuaries. Current president of the Conchological Society and a member of the Council of the Zoological Society of South Africa, he sits on various Government-appointed environmental committees, including that dealing with the conservation of the Langebaan Lagoon. He has many scientific publications to his credit, and his research on limpets is internationally recognised.

Dr Michael A. Cluver provided the material from which the chapter on the Karoo National Park was directly drawn. Presently Assistant Director of the South African Museum, Cape Town, where he has worked as research officer and curator since 1966, he is a Doctor of Zoology whose research interests have centred on the fossil mammal-like reptiles of the Karoo, especially the herbivorous Dicynodontia. He is author of a popular book and a number of scientific papers on the fossil reptiles of South Africa.

Hans van Daalen, in collaboration with Dr Geldenhuys, introduced much fascinating material to the chapter on the Tsitsikamma Forest National Park. Employed at the Saasveld Research Station, George, his study has centred on a continuation of his master's thesis on the dynamics of the forest-fynbos ecotone in the Southern Cape, as well as on the reconstruction of the indigenous forest and increment studies on indigenous tree species.

Dr J.H.M. David was consultant for the Bontebok National Park. His interesting career includes teaching in Nairobi and several years in the computer field. He lectured in the Department of Zoology at the University of Cape Town for nine years and is currently a seal biologist with the Sea Fisheries Institute. His master's thesis dealt with the behaviour of the bontebok, and his doctoral thesis with the population biology of the striped fieldmouse, *Rhabdomys pumilio.*

Dr Gerrit de Graaf prepared the material for the chapter on the Golden Gate Highlands National Park. Following an academic career as lecturer and later as associate professor in zoology at the University of Pretoria, he joined the National Parks Board of Trustees as Liaison and Research Officer. Since 1974 he has been in charge of scientific liaison for the Board.

C.J. Geldenhuys, with Hans van Daalen, provided the fascinating material for the Tsitsikamma Forest National Park. His past work has included the survey and planning of the woodlands in the Kavango and Eastern Caprivi for multiple-use management. Since 1969 he has worked at the Saasveld Forestry Research Station, George, where he is responsible for conservation forestry research with particular emphasis on the indigenous forests, fynbos and associated animal life. The distribution, population biology and dynamics of the Outeniqua yellowwood are the subjects of his master's thesis.

Dr J.H. Grobler, consultant for the chapter on the Mountain Zebra National Park, is a Senior Research Officer for the National Parks Board's Southern Parks. His early life on the Matopos Research Station led to an interest in wildlife and his subsequent doctorate on the roan antelope. Before leaving what was then Rhodesia he worked for nine years for the Department of National Parks and Wildlife Management. He has published widely in various scientific national and international journals, and has also written two guide books and numerous popular articles.

Andrew Gubb, with Dr Liversidge, contributed the material on the Augrabies Falls National Park. Before taking a degree at the University of Cape Town, he was a junior ranger at the Cape of Good Hope Nature Reserve for three years and is now employed at the McGregor Museum, Kimberley, as an ecological botanist and herbarium curator. He is currently involved in the final stages of a book on the large shrubs and trees of the Northern Cape, while at the same time studying for a doctorate on the vegetation types of the Northern Cape.

Dr. Anthony Hall-Martin contributed excellent material for the essay on the Addo Elephant National Park. He currently controls research in all National Parks with the exception of the Kruger National Park, and is the Southern African regional representative on the IUCN/SSC elephant and rhino specialist group. He has worked with elephants in Malawi and Maputoland, and studied them intensively in the Addo Elephant and Kruger National Parks. His other major interest is the black rhinoceros. As a wildlife biologist he has had wide-ranging experience in Malawi, Antarctica, England and South Africa, has co-authored several books and contributed nearly 40 papers to scientific and popular journals.

Dr Richard Liversidge collaborated with Andrew Gubb in providing extremely useful background material on the Augrabies Falls National Park. Director of the McGregor Museum, Kimberley, he is involved in a long-term study of birds in the Kalahari National Park, in springbok, and in the ecology of the Northern Cape. He is associated with several local and international scientific bodies. He has published papers on birds, springbok, grazing grasses and plants, and has co-authored several books on these subjects.

Dr. Gus Mills' research on the Hyaena and its ecology, and on the Kalahari, provided the material on which the chapter on the Kalahari Gemsbok National Park is based. In his capacity as Senior Research Officer for the National Parks Board, he has spent over 10 years working in this Park, during which time his research has focused on the relationship between the ecology of brown and spotted hyaenas and their social behaviour. Other particular fields of interest include carnivores and large herbivores. He is author of 12 scientific papers on birds, large herbivores and lions, as well as many popular articles.

Dr. U. de V. Pienaar, consultant for the chapter on the Kruger National Park, has been associated with this Park since 1955, when he first entered service as a junior ranger. He obtained his doctorate in 1953 with a thesis entitled "The Haemotology of some South African Reptiles." At present he is the Warden of the Kruger National Park. He has represented the National Parks Board internationally, and has published more than 75 scientific papers and books.

Dr. G.A. Robinson, consultant for Tsitsikamma Coastal National Parks was formerly Head of All National Parks, other than Kruger. He now heads the new division of Marine National Parks. He is thus involved in all aspects of park management, from nature conservation to tourist recreation. He has represented South Africa and the National Parks Board abroad and has served on numerous committees. Papers and books that have been published reveal his diverse experience and research, particularly in marine ecology.

Professor Roy Siegfried, in his role as general consultant, supervised the writing of the text and wrote the introductory chapter. A professor in the Faculty of Science at the University of Cape Town, he is also Director of the Percy FitzPatrick Institute of African Ornithology. His professional career has included original research in marine, freshwater and terrestrial ecology, and he is a member of a number of scientific research committees for Antarctica and the Southern Ocean. He has been a visiting professor to the Department of Ecology and Behavioural Biology at the University of Minnesota and a research fellow in Canada and Europe.

Dr P.T. van der Walt checked the manuscript. Since 1972 he has been employed by the National Parks Board of Trustees as a research officer and presently is Control Research Officer in charge of all research in South Africa's national parks, except Kruger. His studies have centred on plant ecology and his Ph.D. in 1973 dealt with the plant ecology of the Boschberg mountain range with special reference to the grazing factor.

P. van Wyk, Head of Research and Information for the National Parks Board, is author of the standard two-volume work "Trees of the Kruger National Park." Before his promotion he was Director of Conservation for the Kruger National Park. His role in the making of this book has been that of general adviser on behalf of the Parks Board. Since graduating with an M.Sc he has been intimately involved in ecological research, and has written numerous articles on aspects of park management and nature conservation. He has represented the National Parks Board overseas on a number of occasions.

---

1. (opposite title page) *Female nyala in Kruger National Park's woodlands at Pafuri.* 2. (overleaf) *Early morning and the rising African sun erases the night from a Lowveld baobab.* 3. (page 8) *Addo's shrouding bush enfolds an elephant bull, revealing why the hunters who pursued the remnants of this southerly breeding herd of African elephants in the early years of this century failed to destroy them entirely.* 4. (pages 10 and 11) *Blue wildebeest charge up the bank of the dry and ancient Auob riverbed in the Kalahari Gemsbok National Park.*

CONTENTS

FOREWORD

In the field of nature conservation South Africa has a long and proud history. As early as 1897 the Hluhluwe and Umfolozi Game Reserves were set aside as the first areas specifically for the conservation of wildlife. Soon afterwards, in 1898, the Sabie Game Reserve was established – largely through the efforts of President Paul Kruger. Its change in legal status from game reserve to national park is the result of the tireless efforts of the first Warden, Col James Stevenson-Hamilton, who finally achieved this goal in 1926. On 31 May of that year, Piet Grobler, then Minister of Lands, announced in Parliament the birth of South Africa's first national park, henceforth to be known as the Kruger National Park in honour of its founding father.

Now, almost 60 years later, Kruger Park has grown from a relatively unknown conservation area into an internationally recognised national park that ranks among the ten most important in the world.

The National Parks Act of 1926 provided for the establishment of other national parks and, since then, nine more have been proclaimed. Several more are in the pipeline. Each encompasses part of South Africa's finest scenery, plant and animal life, and each is a monument to men who transformed their vision into reality, conserving some of the country's greatest natural assets in national parks for all to enjoy.

The question is often asked: Why conserve nature? Why have national parks? The answer, in my view, is simply that without them our lives would be empty and valueless. Nature remains the greatest source of knowledge and abiding inspiration to man. That the national parks and nature reserves fulfil a real need in man's cultural well-being is borne out by the ever-increasing numbers of visitors to these sanctuaries. And without doubt further urbanisation will make more imperative man's escape from built-up areas, continual noise and polluted skies. He will seek in nature, as never before, a truer perspective and spiritual harmony. This need has already reached the stage where at times demands for accommodation in our national parks exceed availability. And because there are limits to the numbers that can be permitted to visit a national park, there is a desperate need for more of the country's ecosystems and natural wonders to be conserved in this way – not only to protect them, but also for man's enjoyment.

In South Africa today natural areas are becoming increasingly rare. The current rate of population growth, increasing demands for food and raw materials, for homes and communication networks, make it only a matter of time before a point of 'no-return' is reached.

While there is still sand in the hour-glass, we owe it to posterity to protect South Africa's unique ecosystems, so that those who follow us may also experience and cherish the wonders of nature. And this must be done now, before it is too late. Dr Rocco Knobel, former Chief Director of National Parks, aptly observed: 'We have to accept that our generation is probably the last that will have the opportunity of setting aside any place of superior scenic beauty, spacious refuges for wildlife, or nature reserves of any significant size and grandeur.'

To achieve our goal of conserving at least a portion of each of South Africa's many ecosystems, public awareness – and support of the system of national parks – must grow and must involve all South Africans. Such widespread public concern and support can best be stimulated and achieved through the written and spoken word and, most importantly of all, by personal experience of nature in all its glory. Everyone is free to visit our national parks, and the more people that do so and come to share an appreciation of them, the more secure the future.

This book portrays in words and pictures something of the splendour and magnificence of South Africa's national parks. While nothing can compare with firsthand experience, it will succeed in its objective if it evokes not only pleasure, but stimulates a desire to visit the national parks and instils a greater public appreciation and understanding of the parks and of the natural systems they contain.

I trust that all lovers of nature will derive as much enjoyment from this book as I have.

A. M. Brynard
CHIEF DIRECTOR OF NATIONAL PARKS

TO HAVE AND TO KEEP

W. ROY SIEGFRIED

THE SUN AND THE MOON AND THE STARS WOULD HAVE DISAPPEARED LONG AGO. . .
HAD THEY HAPPENED TO BE WITHIN REACH OF PREDATORY HUMAN HANDS.
HAVELOCK ELLIS

South Africa has 11 proclaimed national parks; it may have more within a year. . . a decade. . . a century. It may also have fewer, for the continuing success of any conservation movement depends largely on the positive attitudes of society towards it. Without this support, it is conceivable that conservation areas, including national parks, may change greatly and their objectives become very different from those of today; it is even conceivable that without informed concern they may cease to exist.

The national parks attract hundreds of thousands of visitors, mainly South Africans, each year. They come away refreshed and pleased, and most will agree that they enjoy spending their holidays in what they term 'game reserves'. However, few are able to explain what a game reserve really is, or realise that it differs from a national park. Why? The answer is simple: the general public in South Africa is to a large extent uninformed as to what a national park is, how national parks are acquired and maintained, and why they exist at all.

This ignorance can have damaging repercussions, for throughout the world conservation areas such as national parks exist only because society wants them to. It is the public support of the principles involved in the legal proclamation of parks – in South Africa about three million hectares or slightly more than two per cent of the nation's territory – that allows them to exist or brings them into being. Therefore the public should know why and what it is supporting. And it is imperative that this is done, for without this knowledge lodged firmly in the corporate mind, there is no guarantee that

South Africa's national parks will survive into the next century. If this sounds alarmist, it should be borne in mind that there are powerful lobbies working to change parliamentary legislation and so weaken the legal integrity of South Africa's national parks.

Within the embrace of the conservation movement at least ten categories of protection are internationally recognised, with national parks at the pinnacle. These parks are specifically conserved areas of special scenic, historical or scientific interest at sea or on land which are actively protected and whose existence is inviolate. Their legal protection is vested in a nation's highest competent authority. Such inviolate protection is not assured in the case of provincial reserves, conservation areas, game parks and a host of other areas designated for some particular conservation objective and which receive some degree of legal protection. However, in terms of various attributes, particularly their scientific and aesthetic qualities, a number of South Africa's other protected areas are superior to some of the country's national parks; most are open to the public under special conditions that leave no doubt as to the visitor's status as a privileged guest who consequently has responsibilities affecting his behaviour. Usually on entering a park, he is confronted with a set of restrictions – do not feed the animals, do not exceed the speed limit, and so on – and is threatened with severe penalties for breaking the official code of conduct. However, the visitor also needs to be told precisely, and in terms he can easily understand, how the code affects his own best interest and that of the park generally.

The public's right to visit the national parks (and, indeed, most other protected areas) is something that is taken for granted and it has been an element of the national park concept from the start. In 1872, when Yellowstone was established as the world's first national park, the United States Congress included the words: 'as a public park or pleasuring ground for the benefit and enjoyment of the people.' This philosophy is embraced by the 1926 parliamentary legislation which created South Africa's national parks. The law provides for their establishment and for 'the preservation of wild animal life, wild vegetation and objects of ethnological, historical or other scientific interest therein and for matters incidental thereto, in the interest and to the benefit and for the enjoyment of the inhabitants of the Republic of South Africa.' Though this Act provides for a board of trustees to control and manage the parks, it also quite clearly implies that the citizens, through government, should have a say in how their national parks are managed.

And so while it is accepted that the national parks are there to be enjoyed, we must also consider the implications of 'interest' and 'benefit' – whose interest? whose benefit? To find these answers and to understand what national parks are supposed to achieve, we must examine the rationale behind setting aside such areas.

The modern protagonist of conservation in the form of national parks and other conservation areas usually proffers four arguments containing actual or potential benefits to society. But before discussing them we should consider the word

5. *Mature riverine forest along the Luvuvhu River creates a lush backdrop as hippo surge into its cooling waters.*

'benefit' – and the word 'loss'. Benefits and losses to society can be budgeted for and weighed in various ways, but it is important that the same currency be used. Just as oranges and lemons cannot be used as equivalents when weighed against one another, moral arguments, financial considerations and scientific advantages cannot be used to counteract one another. If we are talking about how we feel strongly against culling, it is no use countering the point with how much money would be lost/gained, or what quantity of nutrients would be lost/gained in the process. Furthermore, benefits in cold cash are not the same as economic benefits; economics has far more to it than just money.

The four arguments can be summed up as the ethical, aesthetic, economic and ecological. None exists entirely in isolation and all overlap to a greater or lesser extent.

The ethical argument is simple to understand: it states that man has a moral obligation to preserve the world's wildlife and places of natural beauty. This is a value judgement with which you either agree or disagree. The aesthetic approach boils down to much the same as that put forward to preserve great works of art. Again it is a matter of how you feel. This view also incorporates the idea that we need wilderness areas where we can contemplate our origins and spiritual values while freed of daily stresses.

Both the aesthetic and ethical arguments are generally accepted by first world nations and by many South Africans, but they are likely to be impotent in the face of demands for food by the majority of people living in Africa.

This is especially so where protected natural areas are surrounded by a burgeoning but impoverished population attempting to eke a living from degraded and unproductive land. Though the ethical and aesthetic arguments on behalf of national parks may prove impotent in the future, they are no less valid.

The economic argument usually focuses on tourism: the national parks bring in revenues via tourism and thus justify their existence. Taken in isolation this is a dangerous view. Tourism is not a sufficiently stable base to guarantee the permanency of a park; the argument implies that the park should continue to exist for only as long as it is financially profitable. It is a matter of justifying the long-term ecological benefits versus the perceived loss to the quality of human life now. Since economic benefits are not only monetary, the economic argument is best linked with the ecological one – in my opinion, the strongest of all.

There are four main goals in the ecological argument, yielding four categories of benefits for society: maintaining genetic potentials, monitoring environmental quality, utilizing wild plants and animals, and understanding the functioning of ecosystems.

First there is the potential benefit of a genetic resource. The national parks (indeed, all the protected natural areas) contain plants and animals from which one day it might be possible to develop nutritious crops, medicines of vital importance and domestic animals. Less than 40 years have passed since the range of modern 'miracle' drugs – antibiotics – were developed from substances

produced by micro-organisms. It is thus as important for us to conserve micro-organisms as it is to conserve elephants.

Another important function of national parks is the isolation of 'core' areas which are relatively untouched by man-made pollutants. As a result they provide witness areas – yardsticks against which we can judge the degree of contamination in the rest of our environment.

Plants and animals tend to reproduce as fast as they can, their offspring dispersing from high to low density areas. The animal emigrants can be cropped and their products used to satisfy such human needs as meat, leather and fertilizer. In an ideal situation – were ideal situations to exist – this could be done, but I must sound a caveat: if meat requirements, for example, are to determine how many and what kind of animals will be cropped, we might find that conservation principles are being abused. Too many animals may for instance be taken, thus not only jeopardising the survival of the cropped population but also potentially damaging other species with which it interacts. Thus ecological principles must take first place in every operation of this kind.

Finally, there is the need for ecosystems to function fully and freely within particularly protected areas. 'Ecosystem' was coined in 1936 to describe our emerging concept of the way in which the natural world functions. The concept is not difficult to understand – although its intricacies and consequences are only beginning to be perceived.

The world's plants and animals, in all their number and variety, do not occur randomly but are organised into biological communities in which they

(and mankind) are linked by the flow of energy and by the cycling of chemical nutrients from the soil, air and water. It is this association of living and non-living matter that is called an ecosystem. And because all parts of an ecosystem are linked and interrelated, disturbance of any element is likely to have a ripple effect. The degree depends on the nature of the disturbance and the part which is disturbed; some will have greater effect than others, but time and again we discover the importance of a part of an ecosystem only after the damage has been done.

Interrelationships and their ecological effects are revealed throughout this book. There are salutary lessons and there is the wonder of newly-revealed relationships that change our view of the workings of the natural world and our own place in it. Of course, not only man and his works bring about changes to ecosystems; they are changing all the time, but the effects of man's changes tend to be faster and are often traumatic.

Man's hand is evident in all 'natural' areas, including national parks; they are fenced and protected and the mere fact that so many of these fences cut across natural boundaries, cut off migration routes, change foraging patterns and alter the abundance of animal populations, means that they must be managed. I touch on management here for it shades the meaning of 'natural areas'. National parks and other protected natural areas are undisturbed only by comparison to the land surrounding them. However, they can still yield valuable information that will allow us to discover the workings of ecosystems in relatively untouched environments.

The traditional role of park management was largely a matter of benign neglect. Today managers need to integrate skills and information from a wide field of professional experts – equally the ecologist with his broader views, experts in systems analysis versed in the mathematics of dynamic situations, and the specialist scientist with a narrow field of study. This trend will doubtless accelerate and is in the best interests of the parks system and ourselves.

Now we must turn to the troubling question of how parks should be set aside and maintained. Since 1962 the Parks Board has made it its policy to proclaim as national parks representative areas of each of South Africa's many ecosystems. Prior to that, South Africa's national parks were established for other reasons, among which the desire to preserve the large mammals of the African savannah was paramount. Consequently, examples of many other major ecosystems are represented either inadequately or not at all in the nation's system of parks; the coastline and marine ecosystems have been particularly neglected.

Of the earlier parks many had boundaries that cut across ecosystems rather than contained them. The significance of this has only emerged recently and it is imperative that it be set right, for there is still time for some of the existing boundaries to be reviewed and attempts made to bring them into line with the emerging parameters of ecosystems.

Proclaiming a national park is fraught with difficulties, for as we now realise, it should not simply be an arbitrarily fenced area. South Africa has a limited surface area and within these 1 222 220 square kilometres live some 26 million people now, by the year 2000 it is estimated there will be 50 million in need of food, shelter, work; and their demands for space and on natural resources will be proportionately greater. And so the setting aside of national parks and protected areas becomes a facet of the larger issue of the delicate tightrope between 'benefits' to society and 'losses', between land that will yield immediate returns in tonnes of maize or bales of wool, and land that should be allocated and managed for the inestimably valuable – but somewhat less immediately financially rewarding – needs of conservation.

This was already pointed out some 25 years ago by F. Fraser Darling, a remarkably perceptive ecologist and pragmatic conservationist for his time, in his book *Wildlife in an African Territory*. He commented: '. . . the ecology of land-use is at present in its infancy and should not be equated with the opportunist, profit-seeking, so-called practical approach to apportionment of land-use which is still common', and went on to say: 'Growing understanding of ecological principles raises the conservation of nature to an ethic, the precept of which is that the eternal must not be sacrificed to the expedient.' He called for the 'conservation of the habitat for posterity' and warned that we should 'not . . . concede to current political fads which have no foundation in the ecological principles that govern our ultimate existence.'

National and international concern for the legal status and management of national parks began to develop momentum some 25 years ago, and since then there have been many conferences of experts who have endeavoured to produce a universally acceptable definition of, and management policy for, a 'national park'. This is difficult, if not impossible, not least because our needs and values do not remain constant. Will the 50 million people living in South Africa in 2000 share our views on conservation or will the imperatives of survival make national parks a luxury they feel they cannot afford? Even today, depending on what we do and where we live, our views on conservation differ. A cattle farmer regards a lion as a menace and will destroy it. Yet the urban dweller feels strongly that it should be preserved. So our decision has been to fence special areas where lion roam freely, where we may observe them, and where they cannot prey on livestock.

International criteria governing the selection of areas for national parks are based on three principles. First that protection is entrusted to central government. Secondly, that each park should be of a viable size. A rockpool may encompass an ecosystem, so may a camelthorn in the Kalahari; but ecosystems are worlds within worlds, linked and interlinked. So, just as the protection of a single rockpool will not conserve a length of rocky shore, so a pocket-handkerchief park will not necessarily preserve an ecosystem.

6. (opposite) *Lion at a waterhole in the Kalahari Gemsbok National Park.* **7.** *Too large and heavy to take to the wing, a male ostrich paces swiftly across the Auob riverbed in the Kalahari Gemsbok National Park.*

Throughout this book the problem arises in one or another park, to a greater or lesser extent. It becomes apparent that the Kalahari Gemsbok National Park and Kruger National Park go far towards fulfilling the internationally accepted criteria; they are sufficiently large to permit ecosystems to function. But in some of the smaller parks this criterion is unattainable.

Finally, not only must an area be of sufficient size and be protected by the highest authority, but people should be permitted access to it. These three conditions are hedged, expanded and qualified further in the jargon used by international bureaucracies. In practice, perhaps inevitably, only the first principle specifying the highest legal authority is free of the wide and fuzzy bounds of interpretation that blur the definition between national parks and other categories of protected natural areas.

South Africa's system of national parks is often unfavourably compared with those of other nations in Africa and elsewhere because it accounts for only some two per cent of the country's total landmass. Certain nations considerably smaller than South Africa and lacking her economic strength have proclaimed upwards of 10 per cent of their territories national parks. Admittedly, in some national and international lobbies of ardent and vociferous but not always clear-thinking conservationists nothing short of 10 per cent of a nation's total area is acceptable. However, this magical 10 per cent margin of so-called respectability is entirely arbitrary. It would be very fine if 10 per cent were set aside but, in the final analysis, it is not simply a question of how much land is protected or the number of national parks created; the success of the conservation effort in a country must be judged by whether it achieves its objectives, and here we must consider both management and whether each park is individually large enough to fully protect the ecosystem it encompasses. It boils down to the old adage: it is not only quantity that counts, but quality. Compared to the national parks elsewhere in Africa, South Africa's have an encouraging record in terms of their management.

Management is a thorny issue, but integral to the concept of national parks as a whole. In essence, the management policy should embrace the objective or suite of objectives for which the park or system of parks were set aside. The range of such objectives begins to emerge from this book: for example, there is single-species conservation as practised in the Bontebok National Park; recreational and aesthetic considerations espoused in the Golden Gate National Park; and the much fuller objectives of parks such as Kruger and Kalahari Gemsbok. In all of them some of the stated objectives are being met; some are not, and there is much room for clarification, horizons of thought being widened, opened to public debate and review. A deluge of scientific theory has opened new avenues confronting the men in charge with alternatives and choices on which they should act. The issues are not cut and dried and managers are forced to feel their way.

These are not merely issues for the Parks Board and for government. They are of concern to all of us for we, too, have a responsibility. The public is in the dangerous position of being so ignorant of almost every facet of conservation that it finds itself unarmed against many strongly-held views backed by seemingly-sound arguments. We have a duty to know, so that we understand the issues properly.

This lack of understanding is not helped by media scribes and public pressure groups which pick persistently at superficial scabs and sores, rather than treating the really important causes retarding the development of national parks.

Concern with management arises from the very nature of national parks; none of them, indeed no natural area in Africa, is now large or pristine enough to be left to itself. In the public view, the role of management has tended to have crystallised about two issues. Going back over the last 15 years or so, I have scanned collections of hundreds of reports in daily newspapers concerning South Africa's national parks. Besides the controversy over whether the mining of coal should be permitted in Kruger Park, the two topics that feature time and again are tourism and the killing of, and trade in, ostensibly surplus animals. These, then, are subjects of major, current public concern. The fact that they are interrelated and parts of what should be a matter of much larger concern is not generally recognised.

They basically revolve about management, for both are actions taken to attempt to ameliorate crises that have been caused by the necessary fencing of the national parks. From this first basic act of enclosure spring the myriad other actions and choices that bedevil management today. So often we are attempting to conserve natural systems under unnatural circumstances. And when we modify or change one part of an ecosystem, invariably sooner or later it becomes necessary to modify some other part. Management does beget management, but given the world we live in, there is no alternative. Instead we work towards wise management, and here we depend on tools that are often untried. The chapter on Kruger National Park teases out some of the ramifications as expressed in culling, in the inevitable limits to tourism, and to our status as privileged guests. Management is an art taking its first unsteady steps towards becoming a science; mistakes will be made, but the objectives we chose in setting aside areas for the purpose of conservation remain as beacons should we leave the path.

The proclamation of a national park to represent and conserve each of South Africa's diverse and many ecosystems remains a primary objective. However, since Augrabies was established almost two decades ago, in 1966, there have only been two new acquisitions. Moreover, the Karoo Park owes it existence largely to the initiative of voluntary conservation bodies which encouraged schoolchildren to raise the funds to purchase the land. In many ways the provincial authorities have been more successful in establishing new nature reserves for the conservation of specific ecosystems.

This underscores an important point. With the current drive for rapid expansion in agriculture and industry, the

country's natural areas are being transformed irreversibly as never before. Doubtless pressures will grow in favour of corrupting the integrity of protected natural areas, so sacrificing part of South Africa's natural heritage. The recent public outcry over coal-mining in Kruger National Park gives credence to Fraser Darling's words. We came perilously close to permitting 'the eternal to be sacrificed to the expedient'. If we allow the Act to be changed, the legal protection our national parks have enjoyed thus far may be lost, in which case their continued existence becomes threatened. Any steps taken to enhance the legal safeguards of protected natural areas must be beneficial for nature conservation, albeit that the main benefit might be no more than a temporary reprieve.

Because South Africa needs more national parks it might be prudent to incorporate some of the provincial nature reserves into an expanded system. This could give legal protection to those natural areas which to all intents and purposes are managed as national parks but which currently lack such security.

For it bears reiterating that national parks belong to the people, for the enjoyment and benefit of the people. They belong to you and to me. Part of our taxes goes toward supporting them and acquiring new ones. It is our responsibility to know how the parks are run and where our money goes, and it is our right to demonstrate our concern for their continued welfare – if only for the sake of our children and their children's children. It is worth remembering that we are borrowing something from our descendants, rather than inheriting something from our ancestors. This may be an appropriate note on which to end, but I am not yet finished.

If you have bought this book and have read this far, the chances are that you are well-fed and not short of money. What do national parks mean to poor, hungry people, especially those in Africa attempting to subsist on degraded land surrounding a park? Do they understand why they are prevented from exploiting the grazing for their cattle, and killing wild animals for their cooking pots? Unless they, too, accept the relevance of national parks the future for conservation everywhere in Africa is bleak.

8. *Just emerged from their burrow, a family of suricates soak up the mid-winter sunrise as they warm themselves in preparation for the day's insect-hunting activities.*

THE KRUGER NATIONAL PARK

... THE ETERNAL MUST NOT BE SACRIFICED TO THE EXPEDIENT.
F. FRASER DARLING. WILDLIFE IN AN AFRICAN TERRITORY

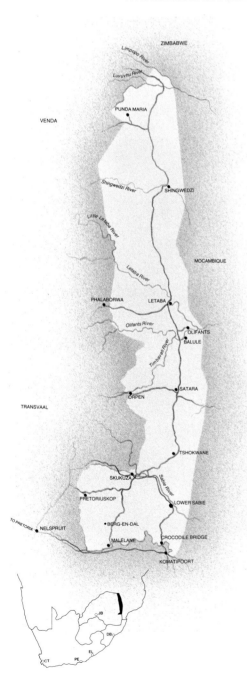

The modern history of conservation in South Africa is linked to a man known in his time more for conservatism than for conservation.

Stephanus Johannes Paulus Kruger, hunter, farmer, soldier and statesman, died nearly 80 years ago, in exile in Switzerland. A sixth-generation Afrikaner of trekboer stock, he was regarded by his adversaries as unbending and gruff, coarse-grained and dour.

To the brash young British empire-builders of the late Victorian era he was quite simply an obstacle to progress and a naive anachronistic relic of the 1836-38 Great Trek. But political opinion is, at best, subjective. History looks back on his era as one of wars, gold and conflict between British imperialism and Afrikaner nationalism. Few mention Kruger's passionate love of the land and of nature, other than to further underscore the theme of political naiveté.

Yet it was this same man, the 'backward-looking' President of the South African Republic, who exhorted his Volksraad colleagues to set aside a tract of land as a sanctuary for game animals. His declaration of intent is a milestone in the story of nature conservation in this country, for although rejected at first, it later led to the establishment in 1894 of the Pongola Reserve.

New ideas need powerful protagonists. Only a man of Kruger's stature and a man whose good faith is beyond question could have proposed that land, good hunting land, be set aside for 'nature'. To the Afrikaner the land has always been central to the national character symbolised, then forged, by the trek ethic,

9. *Kruger National Park's wild northern reaches where the Nyalaland Hiking Trail runs through baobab-studded veld.*

and drawing strength from pragmatic men working close to the earth. Sweet grass and fresh water represent for the herder a paradise on earth; for the farmer a good piece of land can be made to yield rich harvests. Kruger's proposals ran contrary to the mainstream of ideas and sentiment and, predictably, his ideas did not go unopposed. He had to wait until 1898 before he achieved his primary aim – and another milestone in South Africa's conservation story; the establishment of a wildlife reserve in the Transvaal Lowveld. On 26 March of that year he signed the proclamation setting aside the 'Gouvernement Wildtuin' – later known as the Sabie Game Reserve – between the Sabie and Crocodile rivers.

Twenty-eight years later, on 31 May 1926, the National Parks Act was passed by Parliament. Though Kruger was long dead, his dream lived on; he and his fellow conservationists had achieved the next and most telling milestone. In terms of this Act, the Sabie Game Reserve was renamed the Kruger National Park, so honouring the man whose vision and foresight had ensured this unique inheritance. More importantly, the principle of protection of selected areas as national parks was accepted and ratified.

Public acceptance of this idea is no small element in the success of the national parks movement in South Africa. It reflects a largely unanimous view, among white South Africans at least, that

setting aside and protection of areas as national parks is 'good', and that in the eyes of the man in the street, the benefits outweigh the cost. This was dramatically reflected in the recent opposition to plans to mine the considerable coal reserves in Kruger Park. Without this public sentiment, front-end loaders might already be gouging coal where lions snooze replete in the sun.

The idea of what a national park is – and should be – has been greatly modified and refined since those early days. At the time most people saw its main feature as the total ban on the hunting of game within its boundaries. But it was to have other purposes: the world's first such park, Yellowstone in the United States of

America, had been proclaimed in 1872 with one of its stated purposes being 'the pleasuring ground for the enjoyment of the people'.

Paul Kruger expressed a similar desire when he wrote: 'If I do not close this small portion of the Low Veld, our grandchildren will not know what a kudu, an eland or a lion looks like . . .', so fostering ideas which today we tend to accept without question: that land be set aside to conserve wildlife, and that we must be welcomed in these areas as of right.

As a man of his time, Kruger's concern was essentially with the big game animals of southern Africa – not with the snakes, the frogs, the insects and the rodents. The word 'ecosystem' had not yet been coined;

the tangled skein of inter-relationships between plants and animals had yet to be perceived, far less begin to be unravelled, and to most minds 'game' and 'wildlife' were synonymous.

The Kruger Park is the second oldest national park in Africa after the Parc National Albert in Zaire, and in its tale we can discern the gradual development of the nature conservation ethic from the Victorian concept of 'big game' preservation to the present day's holistic approach, in which the whole is seen to comprise many parts of which each is vital to the functioning of the system.

It is a fascinating tale and romantic catalogue of successes and failures, of wise decisions and of mistakes. But the mistakes should not be viewed too harshly, for only now are we reaping the benefits of a century of wildlife research backed by educational and technological advances undreamt of by the early game wardens. And we must ask, what is a mistake? Many of the 'mistakes' have been useful learning experiences. Many of the 'mistakes' have only been such because they did not ensure that the Kruger National Park fulfilled our, the public's, notion of what it ought to be. Many more mistakes will no doubt be made. The one thing we must acknowledge is our still immense ignorance of how ecosystems function.

In its infancy, the Sabie Game Reserve had little to commend it as a place of protection for wild animals or as a place for man. It consisted of some 4 600 square kilometres of unfenced, unguarded Lowveld, riddled with mosquitoes that left men yellowed and weak with malaria, with dreaded bilharzia that saps a man of his strength and his will, and populated with hungry lions. Between 1896 and 1897 it experienced a rinderpest epidemic of frightening proportions and although there has been no major outbreak since then, this disease together with the tsetse fly prior to 1896, added nothing to the Lowveld's appeal.

Sgt Izak Holtzhausen and Const. Paul Bester of the Zuid-Afrikaansche Republiek Police, stationed at Komatipoort and Nelspruit respectively, were expected to keep an eye open for poaching and other infringements of the new law. Even given greater powers – and a clearer vision of their role – they would have found themselves outranked and outmanoeuvred by political events. Two

10. *A cheetah and her cubs rest up during the midday heat. The cubs are born a camouflage grey, becoming spotted at about four months.*

years before the proclamation of the reserve the Jameson Raid had taken place; a year and a half after it came into being, the Anglo-Boer War broke out.

The niceties of 'game reserves' were forgotten. Both Boer and Briton shot what they could for badly-needed rations or for target practice. Holtzhausen and Bester, faced by characters such as the colourful eccentric Ludwig Steinacker, gave way.

Steinacker, a former Prussian army officer, had been commissioned by the British to raise a force of cavalry known as Steinacker's Horse, to patrol the Transvaal/Moçambique boundary. The task suited his personality and he became the freebooting king of Komatipoort, Malelane, Sabie Bridge and the whole Lebombo region south of the Olifants River. But his devil-may-care presence brought a benefit in the person of his adjutant, Major Greenhill-Gardyne. An ardent conservationist, the adjutant took unofficial charge of wildlife welfare in the area and, in spite of the pressures of the warring parties, and the need for food, meted out stern punishment to any poachers he came upon. Few would disagree that he merits the title of first honorary game warden of this area.

When hostilities ended in 1902, the interim government under Lord Milner assigned Major James Stevenson-Hamilton, a British army officer, to take charge of the Sabie Game Reserve which was now reproclaimed. As he himself admitted, he was appointed to the post of 'head ranger' on the strength of having spent the years before the war hunting north of the Zambezi in what is now Zambia. It was to be a temporary post on two-year secondment from his regiment: he stayed for 44 years to become a living legend.

His terms of reference were extremely vague. When he pressed his chief, Sir Godfrey Lagden, for instructions, he was cheerfully directed to: 'Go down there and make yourself thoroughly disagreeable to everyone.' He was uncertain of the extent of his authority and was given no indication of the staff he could recruit or what their pay should be. The only things of which he was certain was that he would have to stop illegal hunting within the reserve and would have to build up the severely depleted animal populations.

Some indication of the enormity of this undertaking became apparent on his arrival; he travelled through the Lowveld for six days before seeing his first sign of wildlife. Conditions were appalling – the animals were nervous and scattered, and poachers were taking the scant remaining game. With peace, men had turned from the pressures of war to the problems of making a living. The Sabie Game Reserve held land with promise: black villagers set fire to its veld to encourage new growth for their cattle; white farmers with a speculative gleam eyed its potential. From the start, Stevenson-Hamilton found himself faced with an obstructive campaign – in the case of the neighbouring farmers and ranchers, an organised one – against his policies.

But Sir Godfrey Lagden had chosen his man well. Stevenson-Hamilton was on one hand an idealist, on the other a man of action possessed of the gifts of organisation and planning. And, perhaps just as important in these fledging years of conservation, he was a superb diplomat and a keen political animal. In his first year he removed squatters, fined poachers, reconnoitred the substantial tract of land under his control, and successfully advocated the disbandment of Steinacker's Horse.

In 1903 he won a vast new world of peculiar wildness and charm to the Sabie Reserve. This, the Shingwedzi Reserve as it was called, consisted of a great swathe of mopane woodland and savannah lying between the Letaba River in the south and the Limpopo in the north. The following year he negotiated with the land-owning and mining companies north of the Sabie River, winning game protection rights over approximately 10 000 square kilometres of land, thus linking the older Sabie Game Reserve with the newly-established Shingwedzi Reserve. He also managed to extend the western boundary of the reserve some 20 kilometres outwards so that more water and better grazing were now incorporated.

In little more than two years he had won protection rights over an area of 37 000 square kilometres, eight times that of the original Sabie Game Reserve. With the exception of the stretch between the Olifants and Letaba rivers, the area encompassed the region from the Limpopo in the north to the Crocodile in the south, and from the foothills of the Drakensberg to the Lebombo range.

The year 1905 saw the start of a new direction in the conservation saga, for Stevenson-Hamilton had become intensely interested in the idea of a national park.

Sir Patrick Duncan was appointed Colonial Secretary and early in his administration the two men discussed the desirability of such a project. Nothing came of the proposal and Stevenson-Hamilton realised that he was unlikely to achieve his aim unless he could enrol public support – and the best way of doing so was to allow visitors into his reserve.

The national park concept Stevenson-Hamilton had in mind was not a refinement of the conservation principles already in practice at Sabie but a concern with long-term protection, which he believed only improved status could provide. He wanted protection at a national rather than a provincial level.

After Union of the provinces of South Africa in 1910, the country entered a period of unprecedented prosperity and development. Stevenson-Hamilton had been able to counter the ambitions of the post-Anglo-Boer War period, but he now came up against more powerful antagonists. The Lowveld was suddenly no longer obscure and unknown, and the companies which had ceded the game control rights in the area between the Olifants and Sabie rivers were not disposed to extend the concessions; there were more profitable uses to which land could be put. Furthermore, on the eve of

11. Immobile and apparently disinterested as they bask, crocodiles have extremely sensitive hearing and can launch themselves into the water with astonishing speed. **12.** *The crocodile's fiercesome teeth are a powerful weapon.*

unification the Transvaal found its coffers overflowing and decided to spend as much of this as it could locally before being obliged to hand it over to the new central government. Part of its last-minute spree was to extend the Selati railway line north from Sabie Bridge to link with Zoekmekaar. The Sabie Game Reserve became a hive of activity as the railway and bridge over the Sabie were built.

Pressure came from all sides. Sheep farmers on the reserve's western boundary agitated for more winter grazing in the Pretoriuskop area. By 1912, despite the Warden's best efforts, 9 000 sheep were bleating in the reserve. For their benefit, again the grass was burnt to encourage fresh growth, and predators were shot on sight. The situation improved only after a grazing levy was imposed.

A classic conflict had arisen between the profit-hungry entrepreneurs and the conservationists working for an environmental principle. It was the time-honoured pitting of sheep against eland; heifers against lion; baboons against mealies. And as this conflict was developing, so World War I intervened and the man who could have won public support on the side of conservation rejoined his old regiment. Without Stevenson-Hamilton and his deputy C. de

Laporte on hand, administration became slack, the opposing camp grew stronger and even agitated against the continued existence of the Sabie Game Reserve.

In 1916, with Stevenson-Hamilton still far away, a commission was appointed to sort out the game reserve issue. Its first report, published in 1918, recommended that the reserve boundaries should not be ratified and that farmers in the Pretoriuskop area should be given winter grazing rights. This meant that the 20-kilometre strip west of the Nsikazi River, taken over by Stevenson-Hamilton in 1903, would be lost to the reserve. This decision was implemented in 1923.

But if conservationists were shocked by these losses, they were delighted by one recommendation made by the commission: that steps should be taken to proclaim the Sabie Game Reserve a national park. Inevitably, nothing was done to implement this recommendation and, in 1921, coal-mining companies began to take an interest in deposits discovered in the Komatipoort area. The Selati railway had proved a financial disaster and the company requested farming rights along the line to help their situation. Matters had reached their lowest ebb. In 1923 the then Secretary of Lands stated that the Department of Lands

intended to subdivide into farms the Crown lands incorporated in the Sabie Game Reserve. To Stevenson-Hamilton, who had returned from war in Europe to do battle on behalf of wildlife, all appeared hopeless.

He desperately needed both popular and political support. He found the latter in the Minister of Lands, Col Denys Reitz, to whom he reiterated his original submission that his beloved game reserve be given national park status. Col Reitz visited the reserve and became committed to the cause.

At about the same time, a tour launched by the South African Railways helped stimulate public awareness of the wilds. The tour's itinerary included the Sabie Game Reserve and it soon became patently clear that the game reserve was the prime attraction. Hundreds of passengers returned home with memories of nights spent under the African sky, of the roar of lions and the many mysterious sounds of the bushveld.

On another front, the Wild Life Protection Society was also agitating for national park status for the reserve. And, inspired by numerous articles in the popular press, wider support had begun to snowball. The idea of a national park gripped public imagination and

Stevenson-Hamilton, assured of government support, sensed the time had come to deal with the problem of private ownership of the tract between the Sabie and Olifants rivers. In an eventual compromise, some ranches were included and others excluded from the reserve.

About 70 privately-owned ranches were expropriated, but the land between the North Sand tributary in the south, and the Klaserie tributary of the Olifants in the north was excluded and deproclaimed. Unfortunately this included some of the best sable and roan habitat in this part of the Lowveld, but its loss was partly offset by the fact that the proposed area would remain a single unit. Today most of those excluded ranches fall within the privately-owned Sabie-Sand, Timbavati and Klaserie game reserves. Another gain was that the proposed national park now incorporated the state-owned land between the Olifants and Letaba rivers.

A change of government at this juncture brought uncertainty. Would it be sympathetic to the national park cause? The conservation lobby's fears were soon allayed, for the new Minister of Lands, Piet Grobler, was Paul Kruger's grand-nephew and shared his concern for wildlife. He assured Stevenson-Hamilton of his full support and promised to devote his energies to seeing the necessary legislation passed.

The birth of the Kruger National Park was not easy and it took all of Grobler's political skills and courage to ensure a safe delivery. Sheep farmers sent a deputation to Parliament condemning the idea; would-be landowners tried to pull strings; and state veterinary officials showed shortsightedness – or at least a surprising lack of insight – in adding their voices in support of slaughter of the game for fear of the tsetse fly returning.

Countering this came a flood of articles and speeches. When the day of reckoning came, Piet Grobler met with the representatives of the land companies. His diplomacy, backed by the threat of expropriation, pushed the pendulum on its swing towards the side of conservation. The land companies agreed to give-up their land within the proposed boundaries in exchange for land outside. At this momentous meeting several hundred thousand hectares changed hands, leaving the Sabie Game Reserve with some 1 900 000 hectares and, but for portions of the western and northern boundaries, little different in shape and extent from today's Kruger National Park.

On 31 May 1926 Grobler introduced the National Parks Act in Parliament. His speech was stirring stuff, presenting the Park as the fulfilment of Paul Kruger's ideal and announcing that henceforth the Sabie Game Reserve would be known as the Kruger National Park of South Africa.

For once the house was in accord. General Jan Smuts, leader of the opposition, seconded the motion. He, too, spoke impressively of the wonders of our wildlife, the necessity to preserve it and the need for national parks as a means to achieve this. Parliament voted unanimously in favour of the national park. The news made headlines. Conservation had won and public sentiment had given its seal of approval.

Many contributed to the establishment of nature conservation in South Africa, but three in particular – Paul Kruger, James Stevenson-Hamilton and Piet Grobler – are its fathers. They represent three phases in its history: the Word, the Deed and the Law.

The Act also determined who would control the national parks. It provided for an Executive Board of Trustees, known as the National Parks Board which to this day is responsible not only for the Kruger National Park but for the national parks that followed and for the ones yet to come.

What were the ideas of these early conservationists? How did the Warden, the Rangers and Board Members see the conservation ethic? From this distance in time it is impossible to say. But we can guess. It was certain to have been relatively simplistic in our terms; an extension of the mediaeval obsession with game animals – essentially those you could dine upon – and those such as elephant and rhinoceros which made for fine hunting photographs and fuelled night-long campfire tales. The early legislation underscores this view, for it refers to certain species as 'royal game', a term redolent of William the Conqueror, Czarist Russia and Imperial Germany. Though Kruger may have wanted his grandchildren to know what a lion looked like, most regarded the king of beasts as something to shoot rather than observe. Stevenson-Hamilton points out in his book *South African Eden* that the 'old' professional hunters had no use for lion as game. The skins had no value and lions were regarded 'merely as vermin, killers of creatures which should rightfully belong to the human predator; when shot, they were generally left where they fell'.

No doubt Stevenson-Hamilton shared this view at the start of his career. In 1902, when he first inspected the Sabie Reserve, he was so distressed by the scarcity of antelope that he decided to adopt a policy of 'carnivore control' as a means of helping the game rebuild its numbers.

In his book (published in 1937 and containing perhaps an element of wisdom after the event) he writes that 'at the beginning it was the policy to keep all carnivorous mammals, reptiles, and to a lesser extent predatory birds, within reasonable limits'. However, he qualifies this: 'having, of course, due regard to the continued existence of every species.'

It is difficult to see why he felt predatory birds should be culled, and although crocodiles do kill 'royal game' when it comes to water, it is far more likely that the warden simply did not like them. His attitude to snakes, even with hindsight, was unambiguous. He writes that black mambas were frequently encountered and usually 'glided with great swiftness up the nearest large tree, where they were easily dispatched with the shotgun I have always carried'.

An unsympathetic view of carnivores is not surprising in a man of Stevenson-Hamilton's upper-class British background. The British landed gentry took game preservation very seriously in the 19th and early 20th centuries and all species of carnivore from weasels and stoats to foxes and birds of prey were eradicated on sight.

And at this early stage in the Park's history such 'carnivore control' was perhaps justified. In 1925, when the national park concept was gaining ground, farmers in the Malelane area protested against the depredations of lions. Stevenson-Hamilton, fearing his reserve might be sacrificed to appease these vociferous agricultural interests, appointed Harold Trollope, a hunter of wide experience and a ranger at Malelane, to 'alleviate the lion menace'. Whereupon Trollope shot out a good number of the predators in the southern portion of the reserve. Stevenson-Hamilton was aware, however, of the dangers of drastically reducing the number of carnivores and on numerous occasions, and in memoranda to the Board, he stressed the vital role of predators in the natural order of things.

It is worth pausing here to examine the wording of the National Parks Act which sets out the aims of the Park. 'The object of the constitution of a park is the establishment, preservation therein of wild animals, marine and plant life and objects of geological, archaeological, historical and ethnological, oceanographic, educational and other scientific interest . . . in such a manner that the area . . . as far as may be and for the benefit and enjoyment of visitors, be retained in its natural state.'

From the start it was not possible to keep the Park in its 'natural state' for at that time it was so depleted of wildlife that the public would have been sorely disappointed. In fact the idea espoused then, and since carried out, has been to create and preserve an area which corresponds to and reflects our idea of 'wild Africa'; one in which as many species of 'big game' as possible are able to survive; and one which we are free to visit and enjoy.

As such the Kruger National Park has been an immense success. Now, some 80 years after its inception, it stands out among the top ten national parks in the world. In Africa it offers one of the finest examples of management of a wildlife area so providing the visitor an unparalleled opportunity.

Although we would like to feel that the animals that inhabit Kruger Park's immense and beautiful wilderness are entirely untouched by man and his machinations, this is no longer possible. The very fact that a national park such as Kruger – or indeed any park anywhere in our over-populated world – is enclosed by humans, is fenced for its own safety, means that it is, in a sense, a human creation.

And yet, because Kruger National Park encompasses nearly two million hectares of African savannah and because this great swathe of grassland and thicket, woodland and river provides such a spectacular array of landscapes and animal life, it is easy to forget the word 'management' and be carried away by the fascination of the creatures and by their truly astounding interactions.

The Park is some 350 kilometres long and about 60 kilometres wide on average. On three sides its boundaries are natural: the Luvuvhu and Limpopo rivers in the north, the Crocodile River in the south and the Lebombo Mountains along the east where the Park abuts Moçambique. Only in the west has the cordoning off of the area by a fence had a marked effect for it cuts across the long-established migration routes of animals such as wildebeest and zebra which, in the past, moved westward each year from the Lowveld in search of better grazing.

The Park has many faces, many moods. In summer, temperatures soar above 40 °C and humidity mounts as cumulo-nimbus clouds jostle and brood above the horizon. The air thickens and the antelope are still, tense, ears twitching, nostrils flared, waiting. Then the rain pelts down. Rivers churn into muddy torrents and that peculiarly African smell of pungent rain-damped earth assails the nostrils.

The images are richly powerful: the animals are sleek and content as they crop

13. *Chacma baboons are often seen in the Park. This youngster will remain in its mother's care, learning from her for up to three years and then must earn its place in the social order of the troop.*

the new season's grasses and nibble on the tender shoots of shrubs. All about the newborn cavort under the watchful eyes of their mothers, for in the long grass lion are drowsing between hunting bouts.

Many visitors prefer Kruger in the winter months of June, July, August when it is dry and the skies are a crisp, cool blue. Because very little rain falls at this time of year, the animals are tense and tend to gather close to the rivers and waterholes. A dry breeze shushes the grass and the long-tail cassias' arm-long pods rattle in its gusts. And, because many of the trees lose their leaves in winter and the grass fades to a brittle gold, the animals are generally easier to spot.

Just as the seasons transform the Park, so each area has a character distinctly its own. Kruger National Park dispels for ever that African stereotype of acacia trees, flat savannahs, dryness, heat and dust. The reality is more complex and exciting.

The Park can be divided loosely into four vegetation regions that reflect, to some extent, both the decrease in rainfall as you travel north through it, and the broad geological division along a north/south axis. On this basis, the western half is granite which has been fairly heavily eroded; the eastern half is essentially basalt. But so greatly simplified an account ignores the special nature of riverine areas and others such as the sandveld regions round Punda Maria in the north-west and along the eastern boundary south of Pafuri which supports a unique community of flowering plants. Indeed the vegetation map of the Park reveals no less than 35 landscapes, each quite distinctive to the trained eye.

Thus the south-west is dominated by an essentially deciduous, large-leafed, woody plant association and tall grasses. South of the Olifants River, which bisects the Park from east to west and serves as a useful reference point when describing the vegetation types, the landscape is largely open savannah in the east with patches of knobthorn, fine marula trees and, along the rivers and dry riverbeds, towering grey-green leadwoods. West of this a mixed combretum-veld dominates. North of the Olifants is essentially mopane country, the trees growing uninterrupted as far as the eye can see, while in the north-east massive baobabs stand sentinel over a landscape of rugged hills and ridges.

But this broad sketch of 'types' is also hardly adequate. From Malelane towards Pretoriuskop and Skukuza camps the landscape's special loveliness owes much to the granite kopjes that thrust up from the plains. Here rock figs send out long roots that hug the boulders and penetrate the fissures. A klipspringer poised like a ballet dancer *en pointe* may appear for a moment in silhouette, statue-like, and then launch itself into a perfect pin-point leap to disappear from view.

In the heat of the day the lizards will be out in force, sunning themselves on the rocks. They do not have the dramatic impact of cheetah on the chase but in their scaled-down world they are equally important as hunters, and play a major role in keeping down the number of insects. Where most lizards are sleek and, like the cheetah, depend on speed for the kill, one lizard has taken the opposite path: the chameleon is slow, stocky and hunts by stealth. The common East African chameleon, the only species found in Kruger Park, is widespread and merits observation. It creeps close to its prey, every movement a study in caution and control. Then, out shoots the sticky prehensile tongue and the unwary grasshopper or mantis is instantly despatched between robust reptilian jaws.

Having opted for stealth, the chameleon is in turn vulnerable to predators such as birds which depend on their eyes to spot prey, and so it has evolved an astonishing ability to change colour to match its surroundings. But in the face of an enemy,

14. *The unappealing swollen stems and sparse leaves of the impala lily give no hint of its spectacular show when in bloom.*

the chameleon summons its full repertoire of threat behaviour, puffing up until it seems twice its size, hissing in menace and gaping to expose its bold red inner mouth. Even human beings, though we know the chameleon can do us no harm, draw back in the face of this overwhelming display.

North of Skukuza, travelling towards the Olifants River, the thickets begin to thin, particularly on the eastern Lebombo flats, giving way to grassy plains. Grass is a major component of the vegetation of the entire Park, but here it is dominant, broken up only by patches of trees. The knobthorn, with its knobbly large-prickled trunk, is common here. This tree is fire resistant and can withstand drought: two factors integral to the development of the African savannah.

Even in Stevenson-Hamilton's day burning the veld was part of management policy, for he realized that to build up the number of antelope he had to stimulate their major food supply. Burning the veld before the rains not only ensures a new mantle for the earth, but together with the effect of browsing animals inhibits the encroachment of bush and trees on grassland. What is more, it has been a determining element in the evolution of the African grasslands for thousands of years. Through the millennia lightning has regularly set the grasslands alight and, ever since their coming, African herdsmen have fired it too, for the same reason that prompted Stevenson-Hamilton. The result has been the evolution of vegetation which is not only adapted to survive regular burning (and grazing) but has, in many cases, become dependent on it in order to survive. This is vividly illustrated when the first rains fall on the charred earth. Within hours it begins to stir. Bulbs burst into brilliant flower and bright green shoots push through the blackened surface. Drawn by this flush, the grazing herds of wildebeest and zebra arrive – as Stevenson-Hamilton expected, and as they still do.

Today the Park is divided into sectors and each year, systematically, some are burnt according to a schedule simulating as closely as possible the established natural fire regime.

Africa is home to some 85 species of antelope, and 20 are found in the Kruger Park. Africa's high preponderance of these creatures can be explained in terms of its vegetation. Kruger Park for example reveals in the variety and extent of its grasslands the numerous niches which the antelope have evolved to fill.

In the Park the antelope and predators take pride of place among the 137 species of mammal. Some, such as the estimated 150 000 impala are almost commonplace; others, like the splendid roan antelope with its coarse reddish-brown coat and swept-back horns, are rare. Only about 350 of this handsome animal occur in the Park which has but a limited area suited to the roan antelope's very particular habitat requirements. However, the resident population is carefully monitored and, because these antelope are prone to anthrax, the rangers inoculate them each year with vaccine, using specially designed hypodermic darts.

Maintaining antelope numbers at certain levels relates directly to how many large predators survive, and their survival satisfies two objectives: firstly, that the Park should contain viable populations of all those animals that lived there in the days of 'Darkest Africa'; and that visitors should experience a sense of wildness, especially as expressed in the hunt.

Lion prefer to hunt in the cool of the night. But if you rise at dawn and go out into the Park as soon as the camp gates open, you may find them still crouched in the grass, a few females deployed to drive the frightened prey towards a lion or lioness lying in wait. The lion in ambush lunges forward, skilfully separates the victim from the herd and brings it down with a quick, clean bite that severs the spinal column. The marvellous economy of action and the hunting strategy bestow a certain dignity on both lion and prey.

Wild dog also hunt co-operatively, the whole pack skilfully selecting and chasing the prey. But when it comes to the *coup de grâce* they lack the lion's sheer power and strong jaws; the kill is slow and traumatic for both victim and human observer, though probably less so for the victim, already in advanced shock.

Like wild dog, cheetah do not have the lion's power when it comes to the kill, and because they generally hunt alone or in pairs they are often unsuccessful, chasing small antelope till the point of exhaustion for the hunter (being sprinters they lack stamina) – and then unable to position themselves for the kill or thwarted by a last-minute desperate change of direction by the prey, they find themselves panting and hungry. Even when successful, they face the prospect of their hard-won meal being snatched from them by bigger cats such as lion, by hyaena, or even by a persistent black-backed jackal. For this reason the cheetah drags its kill to a sheltered spot and there begins feeding immediately on the meaty haunches. Where lion will take their time, assured by their position as king of beasts of a peaceful repast, the cheetah is a

15. *Impala ewe and lamb sniff in mutual recognition.*

hurried, furtive diner who often ends up seeing other animals eat its second course.

Spotting the big cats on the hunt is not common and many visitors to the Park must settle for watching them feed after it is over. Lionesses are usually the hunters, yet they do not get to sample the rewards of their efforts until the males have had their fill. The largest males take pride of place, with the others growling and clubbing one another for a share. After the meal they will lie together in sociable repleteness, but at table it is each for himself. Only when the males are sated do the females settle down to feed, and only if there is enough will the youngsters who are weaned but have not yet earned a place in the dominance hierarchy satisfy their hunger. The small cubs suckle and are thus provided for, though when times are hard they may well find their mothers exhausted and the supply of milk limited.

The relationship between the availability of preferred prey in Kruger Park – impala, wildebeest, waterbuck and zebra – and the well-being of the big predators such as lion, cheetah, leopard, wild dog and hyaena, is not as one would expect. It has been shown that the number of prey animals sets an upper limit on the number of these predators that survive, and not the other way round as we would presume. In fact predators do not usually threaten the survival of the prey, for they do not normally feed off the fittest individuals in the herd, nor are they able to jeopardise the breeding core; they pick off the stragglers. Put in financial terms, they feed off the surplus but do not affect the capital. If anything, they have a positive effect, for in hunting down the weakest members, the sick and the careless, they improve the stock by leaving only the strongest and most viable to breed.

Kruger Park experiences dry and wet weather cycles of about eight to ten years each, though the wet cycles tend to have longer effect, their influence softening the first of the dry years. Years of drought, such as were experienced in the early 1960s, clearly illustrate the way in which the antelope population accounts for how many lion there are. In those disastrously dry years lack of food and water diminished the antelope, leaving them weakened and vulnerable to disease.

Initially the lion had a field day, picking off the old, the young, the ill and disabled. So great was the glut that even the hyaena and vultures were well-filled. But as the drought dragged on, the lion began to suffer: there were now fewer potential victims, the herds were nervous and more alert to approaching danger. A strong impala in full flight can outdistance a lion; a lion is no match in terms of speed for an adult zebra or even a wildebeest. Hungry and tired, the lion were now themselves vulnerable to disease. Many of the females failed to breed and those cubs that were born died of lack of food, and from disease. The sub-adults also fared poorly, for they were still too small to demand a place at the kill. Mewling and desperate they watched and waited. Occasionally, a youngster driven by hunger attempted to snatch a morsel – to find itself rolling in the dust from the mighty clout of a full-grown paw. Many died from such injuries, many more fell ill and did not survive. When the drought finally broke, the lion population had fallen significantly; the antelope numbers had fallen, too, but within a few good seasons most species had recovered.

The rainy season is the season of birth, particularly for the plains animals such as impala, for the zebra and for that comical antelope, the wildebeest. The birthing

period is synchronised over a matter of a few weeks. Quite suddenly the females go into labour; day after day the young flop from the womb, their gawky legs still entangled in the foetal sacs. Within minutes they are taking their first shaky steps, their anxious mothers nudging them on. This is a very vulnerable moment for mother and calf. They are usually apart from the main herd and without the benefit of its collective warning system of dozens of alert eyes and ears. Yet in these few precious, dangerous seconds the mother and young imprint their personal signature – their smells and voice – upon one another. And now, some 10 minutes into this world, the calf on long legs that grow nimbler with every step, is introduced to the herd as it moves on. Although there may be 200 animals in the herd and a third again of them may be newborn, every mother knows her infant's call, its smell, and it, in turn, recognises hers. This bond ensures that she suckles her own infant and that it knows to keep near her and safe.

This sense of safety is easily shattered, for the birthing season of the plains animals marks the beginning of the predators' great annual feast. Summer visitors to Kruger Park experience a morbid fascination and strange chill of fear as the predators plunder the newborn, but soon these feelings are replaced by a sense of awe at the beauty of survival. A lion can only eat so much. At a certain point a cheetah's sleek frame becomes uncomfortably distended. The wild dogs are tight-bellied and satiated. Even the hyaena must reach a point where it can eat no more. And so, after the initial blood-letting, a calm settles over the herds. Danger has passed for the moment and a certain harmony takes its place: some have been killed but many more will now survive.

North of the Olifants River the character of the Park changes yet again. This is (except in the west and extreme eastern sections) the southernmost limit of the mopane, which grows in unbroken stands that seem to stretch forever. It was country like this that the early white hunters referred to as 'MMBA' – 'miles and miles of bloody Africa.' There is often a special silence in the grass-filled spaces between the trees, and at midday the butterfly-shaped leaves turn edge on to the sun and cast very little shade. In summer the cicadas take up residence and the mopane veld rings with their strident call. When they pause the silence is startling until broken once more by the hoarse clatter of hornbills, the shriek of a rednecked francolin or the dry scratching of a flock of crowned guinea-fowl foraging the ground. And then, quite suddenly, there is a rustle and the apparent emptiness of the mopane veld is filled by its most prominent citizens. A herd of elephant comes into view, trunks stripping great mouthfuls of nutritious mopane leaves.

An adult bull elephant of 5 000 kilo-grams needs as much as 300 kilograms of food in 24 hours: thus the requirements of a herd of 30 elephant are immense and, when one considers that today there are more than 7 000 in the Park, their tremendous impact on its vegetation is not surprising. As John Hanks writes of the Kenyan herds in his book *The Elelphant Problem* 'elephants are incredibly careless and extravagant in their feeding habits, well deserving their reputation as the animal bulldozers of the African bush'. The elephants in Kruger Park are equally profligate. Furthermore, in its sanctuary they have no natural enemies and the populations have grown accordingly.

Ivory was the hunters' prize in the early days and they shot out the elephant. When Stevenson-Hamilton took control of

16. (previous page) *In a whirl of dust, eland accompanied by a magnificent kudu bull race across the sandveld in the north of the Park. Attempts to domesticate this hefty (700 kilogram) beast have been partly successful.* **17.** *Both the fever tree and the malaria-bearing anopheles mosquito favour periodically inundated habitats such as this depression which floods when the nearby Limpopo River is in spate.*

the area later to become Kruger National Park he was appalled to find no more than a handful of elephant remaining. Many had fled to Moçambique. The rest had fallen to hunters and the many poachers who entered the Sabie Game Reserve to continue the highly lucrative business of ivory trading.

Stevenson-Hamilton dealt firmly with poachers from the start, but as far as the elephant was concerned it seemed that protection had come too late. But one of the miracles of nature is the speed with which it recovers from disaster: Stevenson-Hamilton's handful, augmented by elephants which returned to the Lowveld from Moçambique once the guns had been silenced, had by 1946 reached an estimated 500. In 1959, a survey made for the first time from a light aircraft revealed the happy news that the population had doubled to 986. In the same year fencing the Park began, preventing animals from filtering into areas outside the boundaries when they became too populous within the Park itself. A 1964 aerial count showed a clear trend: protected from man and limited to the Park, the elephant population was growing at an alarming rate. It then stood at 2 374. A mathematical estimate of the elephant population of Kruger Park today, had it been allowed to grow unchecked, is more than 20 000.

The issue of the carrying capacity of the Park with regard to elephants relates equally to its carrying capacity for humans. How many people can visit and be accommodated overnight in the Park without becoming intrusive and damaging? Plans are afoot to increase the number of beds available by a further thousand. In 1981 almost half a million people visited Kruger National Park: to provide them with the basic amenities – with water, food, fuel and sanitation – takes up an increasing area and drains resources that might well be used by wildlife. This multitude of tourists issues out each day along the 750 kilometres of tar and 1 200 kilometres of gravel road to try their luck and see as much as possible of the Park. These roads undoubtedly have an impact – and not only during construction. For example, when rain falls it runs off the roads and collects on the verges. Encouraged by the increase in moisture, the grasses grow tall and vigorous, the shrubs become ever more expansive. But this verdant fringe also blocks out the visitors' views and so it must be cut back and cleared by hand.

There will undoubtedly come a time in the near future when the Park managers will announce that the limit to human

18. *Within the vast sanctuary of the Kruger National Park animal populations have burgeoned so that today species such as the impala shown here are so numerous (an estimated 150 000) that they must be culled from time to time.*

utilization has been reached. Perhaps many visitors will have to stay in accommodation outside the Park's boundaries and enter the gates on a strictly controlled basis in terms of daily numbers. How will the public react? Will the future stability and conservation principles take precedence over the public relations aspect of this magnificent Park? It is a troubling question. The answer is unlikely to be popular but it will certainly be necessary – as necessary as control of the number of elephants.

The elephant's dining habits – which include pushing over fully grown trees for the sake of a titbit of greenery from the crown, and stripping bark leaving trees vulnerable to disease and pests – combined with the pressure of such growing numbers, doubtless would have led to severe destruction of the habitat, not only for elephant but for the many other creatures that share it. Allowed to run unchecked, the situation may well have emulated the horror of Tsavo East in Kenya where thousands of elephants died of starvation in a denuded and degraded landscape.

In 1965, after it became apparent that the elephant were becoming a threat to themselves and to their habitat, a decision was taken to reduce the population and maintain it at a level which the Park managers felt the environment could adequately sustain, yet still assured visitors of a good chance of observing these imposing beasts. In 1966 culling of elephant and buffalo began. (The latter had also become very abundant – more than 30 000 today – after being all but

eradicated by rinderpest in 1896.)

Culling has raised more emotions and outbursts of public opinion than almost any other aspect of management in South Africa's national parks. There are powerful arguments on both sides and most combine a subjective view with a technical prediction. But we have to admit that we do not know in the long-run whether culling will have any seriously damaging side-effects. Some potential dangers have already been pointed out. One theory advanced is that culling may lead to the evolution of a new equilibrium between animals and plants. This theory claims that culling, by removing nutrients in the form of the carcasses from the environment, may in the long term lead to a lower level of vegetative production which will, in turn, be able to support fewer animals. Eventually plants and animals may reach an equilibrium at levels sufficiently low to remove the pressures that presently prompt culling.

The antagonists' most reiterated objection is that culling is morally repugnant and that it violates the spirit under which conservation areas were proclaimed. There is no answer to these charges for they are not rational but emotional. Some believe that culling may undermine public support for the national park movement and thus weaken its position. Then there are those who believe it best for nature to 'take its course'. Since we do not know exactly what that course might be and because within the confines of a national park 'nature' is already being restrained to some degree from 'its course', this is in

19. *Bare-headed to limit messiness as it delves into carcasses, the scavenging marabou wields a powerful and awesome beak.*

fact a radical suggestion. On evidence from elsewhere in Africa, 'nature's course' may result in a very different Kruger Park and possibly one unacceptable to us. Indeed, the whole matter of what is acceptable lies at the heart of the culling/non-culling issue. There is no single answer, there is no right answer.

Since one of the purposes of Kruger National Park is to maintain viable populations of all those species of animals that lived there within the last few thousand years, and equally to maintain the varied communities of plants and the character of the landscape, it would seem that the suggestion of letting nature take its course is an extremely risky alternative.

Indeed, culling in Kruger Park appears to be the conservative and moderate option, provided that its effects are monitored and the system remains flexible enough to be stopped at any time if management objectives are not being met. It must be added that, while evolution is inherently unstable in the sense that it is dynamic and constantly changing, in the Park such instability is regarded as undesirable and unacceptable. If we acknowledge with honesty that we would like to see the Kruger National Park continue as it is, then culling takes on a different perspective.

Once the decision to cull had been taken, the next hurdle was technical. How should it be done and what should be done with the carcasses. Here Kruger Park

has an impeccable record. Faced by the public furore over culling, the Park managers invited, amongst others, the Wildlife Society of Southern Africa, the South African Veterinary Association, and the South African Federation of Societies for the Prevention of Cruelty to Animals and Affiliated Societies in association with the International Society for the Protection of Animals to visit the Park and observe the culling operation. All these recognised bodies issued statements to the effect that the methods being used were as humane as currently possible.

Elephant and buffalo are brought down with hypodermic darts specially developed at Kruger Park (and since used widely elsewhere in the world). The darts, filled with scoline, a drug which is a muscle paralysing agent, are fired from helicopters. Destroying individuals within a family group of elephants creates immense social upheaval and so entire family groups are selected for culling. The operation is quick, the elephants being immobilized by the scoline and then immediately despatched by a bullet to the brain. Buffalo are rapidly killed by the effects of the drug overdose.

The elephant and buffalo carcasses are loaded onto special trailers and taken by road to the processing plant which is in the Park but is not accessible to tourists. Indeed the entire culling operation is kept from the public eye, simply because it distresses so many and because it presents a somewhat visceral aspect of Park management. The carcasses are processed and the products are sold, raising revenue which is reinvested in the Park. Notably, the scientists who suggest when and where culling is necessary and who put forward the numbers and species (for at times other animals such as hippo and impala have been involved) to be culled have no say in the marketing of the products or in the use to which the profit is put, thus ensuring that their judgement is not swayed by financial considerations.

Because buffalo in the Park are known to be carriers of foot-and-mouth disease, particular care has to be taken in the processing of the carcasses and so far no outbreak of the disease outside the Park has been attributable to the culling programme. Where culled animals are found to be diseased they are left in the veld and nature's garbage men – hyaena, marabou and vultures – rally to the task of cleaning up.

The aesthetic element in culling is perfectly illustrated by the preservation of the big tuskers, bulls such as the famous Mafunyane – 'the irritable one'. The

public enjoys seeing these great specimens with their splendid ivory. Mafunyane lives in the far north of the Park, his whereabouts kept secret for his own safety. Although too old to breed, he and several other bulls with fine tusks are specifically excluded from the culling programme. Mafunyane has a perfectly matched pair of tusks each three metres long and angled so that he is forced to carry his head high to accommodate them. His value to poachers is undoubted and on his head he bears the bullet scar from one such encounter.

Today the elephant population is kept to a figure of about 7 500. It is interesting, though, to note that for the last 500 to 1 000 years at least it is unlikely that the Lowveld supported elephant populations even this high. Evidence for this is the excellent condition of the many baobabs in the north of the Park. Elephant have a predilection for these trees, stripping the bark, eating the fruit and leaves and often leaving the baobab so damaged that it collapses in a great pulpy heap. It is unlikely that so many large baobabs would have survived unscathed had there been many elephants about.

Baobabs distinguish the northern reaches of the Park. Here, before it reaches the Pafuri region, the Luvuvhu River courses through a deep and rugged gorge with thickly wooded sides. This area was only recently opened to tourists. The Nyalaland Trail, one of three wilderness trails in the Park, allows visitors to experience on foot the peculiar charm of this part of the Lowveld. Accompanied by a trained conservator and rangers, parties of eight tourists walk the veld, learning at first hand about the plants and creatures they encounter. They may be lucky enough to see the nyala, often confused with the kudu but in fact quite distinctive in its smaller size, white-tipped lyre-shaped horns on the males, and its yellow-orange knee-sock markings. It is found in abundance in the Luvuvhu's riverine forest. Other relatively rare antelope in the northern half of the Park are the eland, tsessebe and roan antelope as well as the sable antelope, which must rank with the gemsbok as one of the handsomest of Africa's larger antelope species. It is also home to the tiny and secretive Livingstone's suni and to the equally elusive Sharpe's grysbok.

It has been the policy over the years to reintroduce to the Park all such species that originally lived in the Lowveld, but which had been shot out or scattered. The black rhino had last been seen in the Eastern Transvaal in 1936 and presumably had been hunted by poachers keen to obtain rhino horn which is neither horn (it is made up of compacted hairs) nor an aphrodisiac as popularly believed. In 1971, 20 of these ill-tempered beasts were brought in from Natal and subsequently more arrived. They settled down and have bred so that today Kruger Park has about 100 black rhino.

The somewhat larger and gentler-tempered white (square-lipped) rhino was also reintroduced from Natal in 1961, having been killed off in the Kruger Park years before its proclamation. The black rhino is a browser, using its pointed upper lip to pluck shoots from bushes, but the white rhino is a grazer, cropping and plucking up grass with its wide mouth. Indeed its name comes not from its colour – it is a dark dusty grey – but from 'wide' referring to it lips. Seven hundred now inhabit Kruger, mainly in the south in the area between Pretoriuskop, Malelane and Skukuza.

And, just as rhino and various other species such as oribi, grey rhebuck and red duiker have been brought into the Park, so animals which are numerous and for which there is a demand elsewhere are translocated. One of the arguments against culling is that animals should be sold or translocated rather than killed. But this is not always possible. Disease prevents translocation of buffalo, for example, as they are carriers of illnesses dangerous to domestic stock. Elephant are in over-supply to zoos and wildlife reserves in most parts of the world. However, as far as possible this alternative is sought by the Park managers before resorting to the scoline dart.

Hippopotamuses occur in all the Park's perennial rivers, yawning and grumbling the daylight hours away and coming out at night to graze on the riverbanks. Like elephant, they have few significant natural enemies other than man and their numbers have increased to the point where they, too, are culled from time to time. They share the waters with the crocodile, the two living in mutual acceptance. Crocodiles are cunning and dangerous and have killed numbers of people in the Park, especially along the boundary rivers.

Kruger lore is full of tales of lucky escapes, but these are matched by the silence of those who can today tell no tales.

Fishing is forbidden and so the visitor is unlikely to see the large-mouthed bream which is endemic to the Sabie River, or the several other equally interesting and unique fish such as the lungfish and the brilliantly-coloured killifish. However frogs, fish and terrapin are conserved with

20. *A stallion mounts a female Burchell's zebra as the others vocalize and jostle.*

the same attention as the more easily visible species, and the case for the damming of the Park's perennial rivers rests in part on their needs.

Water has long been a rallying point of public interest in Kruger Park's history. Fencing of the area certainly imposed new stresses and restrictions on animals which had previously moved away from the Lowveld in times of drought. The animals were now deprived of this option and the rivers, particularly in dry years, provided only limited relief. Early on the decision was taken to augment the supply by sinking boreholes. These were to be of dual benefit. The ever-increasing flow of tourists required purified water, and the boreholes could provide an alternative source for thirsty herds no longer able to migrate.

In the late 1920s and early 1930s the Park was gripped by drought. The plight of the animals roused great public sympathy and the 'water-for-game' campaign got under way after a Mr Cloete bequeathed money for the erection of the first windmill. Companies as well as individuals provided funds and by the end of 1933 two drills were hard at work. The public had found a cause that appeared to benefit the animals and at the same time allowed people to make a

personal contribution to the conservation effort.

The water-for-game programme is an ongoing one and very costly in the case of major dams. It is also a programme which has been challenged in some quarters as an unwanted interference in natural cycles of drought and abundance. The current policy, however, is to compensate for the ever-increasing amount of water drawn from the rivers before they reach the Park and to stabilize the water supply to ensure the survival of the Park's wildlife even in times of severe drought.

Research in the Park continues at Skukuza, revealing new and wonderful interactions between animals and plants. And as the data comes in, as the statistics emerge, a fuller picture begins to take shape, allowing the Park managers to adjust and adapt their management policies. They seek a degree of stability in an inherently dynamic situation and their task is thus difficult. Their success to date is in no small part tribute to flexibility. Survival in the animal world is characterized by this attribute. As the most successful animals of them all, we are innately flexible and so the foreseeable future of the Kruger National Park and its immeasurable wealth of animal and plant-life is assured.

21

21. The crowned guineafowl occurs in conspicuous flocks that may number hundreds of noisy individuals, scuffling and pecking for seeds or bulbs and insects or worms.   **22.** Culling, one of the most emotional public issues in national parks, extends to hippo. With few natural enemies, they have increased rapidly and great numbers fill waterways where they yawn and grumble the daylight hours away. At night they emerge to graze the grassy banks, their cropping encouraging fresh growth.
23. Much rarer than its cousin (21), the crested guineafowl tends to inhabit the forest fringe, rarely emerging from its woodland home into the open.

22   23

24. *The suni, rarest antelope in the Park, is confined to woodland where it browses often singly, but sometimes in pairs or small groups.* **25.** *A sinuous-necked darter dries off after fishing in the Luvuvhu.* **26.** *Woodstorks in the Luvuvhu. These birds have been seen to steal fish from the mouths of crocodiles which must juggle their catch into position for swallowing.* **27.** *Solitary, secretive, nocturnal and thus not usually seen, leopards are widespread in the Park.*

24

25

26

28

28. *A warthog heads down through a grove of immense sycamore figs to the nearby Luvuvhu River.*   29. *Shortly before the photograph was taken, this martial eagle had killed and eaten the guineafowl which now distends its crop.*   30. *Largely restricted to the northern*

regions of the Park, nyala are nevertheless fairly common along river-banks. The males are striking in yellow-orange knee-sock markings and a dashing spinal crest which they erect when approaching other males. The females are more modest in colour and size.*

29

30

31. *A breeding herd of mothers and young almost dwarfed by their setting. The cow at the head is a matriarch who, by virtue of age and experience, is the dominant member of the group.* **32.** *Vervet monkeys are ubiquitous in the Park.* **33.** *The legendary Mafunyane reveals his famous three-metre long tusks that rank as the most massive on any living elephant. Such superb tusks are an obvious target for poachers and his whereabouts is now a closely-guarded secret, but his temple bears conspicuous evidence of an earlier confrontation with a poacher's bullet.* **34.** *Big Haaktand (Hook tusk), another of Kruger's outstanding bulls had some of the longest tusks of any elephant, until he died recently. His left tusk was 3,17 metres long and weighed just under 53 kilograms; the righthand tusk was only slightly smaller. Such well-endowed bulls are specifically excluded from culling operations.*

31

32

33

34 ▶

36

35   38

37

35. A Cape buffalo bull tolerates the red-billed ox-pecker riding the mighty boss of its horns, for the bird picks ticks and flies off its bovine host. **36.** The dwarf mongoose is the smallest member of the African mongoose family and a fairly common sight in the Park. **37.** A typical scene encountered by many visitors to the Park as a kudu bounds across the road. **38.** On the savannahs of the central areas of the Kruger Park plains animals such as these buffalo are typical and numerous. **39.** (overleaf) Along the Sabie River on the eastern boundary of the Park a tree euphorbia stands sentinel over a typical Lowveld landscape.

40

41

40. The stillness of Lower Sabie's riverside is broken by the harsh clack, clack, clack of marabous squabbling. These birds include carrion in their diet, and even vultures give way when they arrive at a kill.   41. Drinking is a vulnerable time for antelope and these impala had approached the water's edge with utmost caution. The click of the shutter was enough to send them bounding away.   42. As it hunts lizards, a spotted bush snake blends beautifully with the foliage of a lowveld cabbage tree. Most snakes are harmless to man and this is no exception.

42

43

44

43. The small and jewel-like painted reed frog makes a vociferous contribution to the night-time orchestra of the bush in summer. Only the males call, their penetrating whistles indicative of the immense numbers of these tiny creatures.   44. A charaxes butterfly and paper wasp share a sip of sap.   45. A minute grasshopper hitches a ride on a flap-necked chameleon which, gingerly and holding itself high off the hot surface, crosses a tarred road.

45

46

47

48

46. *This female giraffe was in oestrus and a male sniff checks her readiness to breed. Giraffe are found only in Africa. Their tremendously long necks have evolved literally to lift the giraffe above the heavy competition for browse and graze closer to the ground. Though, like those of other mammals,* *its neck contains only seven vertebrae, these bones are enormously lengthened.* **47.** *When Kruger National Park was proclaimed it contained only a relatively small portion of habitat suited to the roan antelope's highly specialised feeding requirements. Only about 350 live there today, and their well-being is of* *great concern to the Park managers for roan antelope are prone to anthrax and so, each year, rangers in a helicopter using specially-designed hypodermic darts inoculate the herds.* **48.** *Impala males battle for dominance which wins the right to breed with the females within a particular area.*

49. *Clamouring and demanding, a lappet-faced vulture raises its wings as it asserts its dominance at an impala carcass. The mill of other vultures – Cape and white-backed species are visible – give way, for the lappet-faced is the largest of the vultures and thus takes precedence in the pecking order.*

50. *Years before Kruger National Park was proclaimed, hunters destroyed the last white rhino in the area. Its horns were, and still are, avidly sought as an aphrodisiac in the ill-founded oriental belief in its power to restore virility. The future of the huge animals remains bleak outside protected areas. In 1961 the white rhino (white bears no relation to its colour but is a corruption of the Dutch for 'wide', referring to its broad lips used for cropping and plucking grass) was reintroduced from Natal and several hundred now inhabit the southern areas of the Park. The so-called black rhino – distinguished by its pointed upper lip used for browsing, and by its cantankerous disposition – was also introduced and today has multiplied to more than a hundred.*

51

52

51

51. From her elevated vantage-point, a lioness watches her potential next meal – a wildebeest grazing beside a nearby pan.   52. Drought is a cyclical threat to the Park; approximately eight to ten years of good rains being followed by a similar number when the rains are poor or even fail. At such times the wildlife suffers acutely and over the past 50 years many boreholes have been sunk and rivers dammed to allay the effects. The boreholes attract animals, and visitors who sit quietly and patiently near these water-points are likely to be rewarded by sights such as these lionesses lapping their fill.   53. The pride's dominant male mates with one of the females. Just over 100 days later she will give birth to two to four cubs which will remain with her until they are about 18 months old – if they survive disease, hunger and the exigencies of being young and helpless in the stern hierarchy of the pride. 54. A post-prandial yawn from a fully-grown male.   55. (following page) Spots and stripes at a waterhole tableau in Kruger National Park.

54

THE TSITSIKAMMA COASTAL NATIONAL PARK

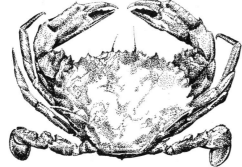

BY THE DEEP SEA, AND MUSIC IN ITS ROAR;
I LOVE NOT MAN THE LESS, BUT NATURE MORE.
BYRON. CHILDE HAROLD'S PILGRIMAGE IV CLXXVIII

The Tsitsikamma Coastal National Park is one of several national parks which invite the visitor to explore and experience it at first hand. Here you can walk trails that take you along this 67-kilometre stretch of magnificent southern coast; watch, entranced, the strange and fascinating world within a rockpool; ponder at the middens left by Strandlopers of long ago; and stroll at leisure through fynbos and forest. And always nearby is the sea. On South Africa's southern coast the warm Agulhas current swings offshore and cooler patches upwell inshore creating an intermediate zone. It nurtures a singular marine environment in which tropical fish and invertebrates can be found – often fewer and smaller than their warmer water brethren – side by side with creatures associated with the cold west coast. Then, too, there are the intriguing sea creatures found nowhere else but along this coast.

None of this is apparent to the casual visitor. The first impression is of a wondrously wild coastline backed by fynbos-clad slopes and cliffs and cut by sheer gorges through which amber-coloured rivers run down to the sea. In these gorges and beyond the fynbos the climax forest reaches imposing and tangled to the Tsitsikamma Mountains.

This coastline is extremely rugged, the sea rarely calm. The 6 500 people who each year stretch their muscles and free their minds by tackling the five-day Otter Trail along the Park's length are generally more struck by their experience of the land than by the sea, although dolphins and whales, sea-birds and clawless otters are memorable if you spot them. Yet the reason for proclaiming this 1,6 kilometre-wide strip of coastline together with the waters for 800 metres offshore is essentially marine not terrestrial. The true wonder of this National Park lies at the ocean's edge and below.

At the Storm's River mouth there is the opportunity to join a trained conservator, don flippers and snorkel, and take to the water to experience South Africa's first underwater trail. With expert guidance the visitor is introduced to the animals and plants of this coast, to their amazing life-cycles and behavioural adaptations. But most visitors – day-trippers as well as those spending a longer period in one of the delightful log cabins – wait for the changing tides to reveal something of this underwater world.

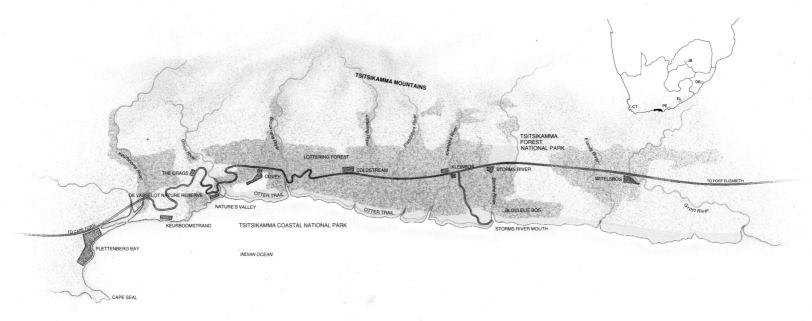

56. The infratidal zone, revealed by low-water spring tide and encrusted with glorious purple soft corals and dark, leathery-clad sea squirts. Two, at centre left and right, can be seen jetting water.

57. *Drawn by the odours of death, plough snails converge on a stranded sea snake. These marine reptiles are related to the equally venomous cobras, and range the warm oceans of the world, feeding on small fish.*

The tides change twice every 25 hours, giving two low and two high tides during this time. Once a fortnight, the day after full and new moon, because of the moon's gravitational pull combined to some extent with that of the sun on the earth's oceans, a spring tide occurs. Then the sea rises higher and recedes further than usual. Spring low tides are ideal for exploring the rocky intertidal area, for they expose the bands into which plants and animals are zoned, largely because of competition between species.

But another more obvious factor that determines which creatures flourish along the Tsitsikamma coast is the immense power of the waves, built up as they race thousands of kilometres before dashing themselves against these shores. And because Tsitsikamma's coast plunges down to the deep sea-bed without the benefit of a gently sloping continental shelf, the waves' fury is not tamed. Given the impressive size and vigour of the seas, the typical zonation of the southern coast is modified here to accommodate creatures and plants adapted to rough water.

Highest on the shore is the splash zone where few marine animals venture. It is too dry and harsh an environment for them although certain crabs make this their home.

The first truly marine zone is the Littorina zone, high on the shore but reached by high tides. Here lives a single marine species of snail, the tiny *Littorina africana*, its population creating a thousand little 'beads' on the rocks. One or two algae also survive here but the Littorina zone is not particularly rewarding.

Far more exciting to the observer is the Upper Balanoid – the barnacle – zone which is bathed by high tide and thus a somewhat less stressful habitat. The creatures that live here have to cope with daytime heat and water-loss but can count on being submerged by the sea for at least half their lives. There are a number of molluscs (recognisable to most people as animals with shells), as well as a great variety of other invertebrates such as sea urchins, starfish and sea anemones to enchant rockpool enthusiasts.

Another intriguing inhabitant here is the shore spider, of which there are two species. Although not truly marine, they have evolved remarkable means by which they can remain underwater for up to 12 hours.

One of the devices they use is easy to see, for these spiders spin a tube-like silken home in a small cleft in the rock and seal it with silk. Behind the silken seal enough air is trapped for the spider to breathe and for its eggs to be protected from the sea. However, Bruno Lamoral, one of South Africa's leading 'spider men', observed that the sea spiders' adaptations go further than this; they are covered with fine hairs coated in a water-repellant substance and, if caught under the sea by the changing tides, this coating creates an air bubble about the spider. But how can the spider survive on this tiny amount of air for so long? Lamoral discovered that the bubble is like a lung; it allows oxygen in the sea water to diffuse into the bubble while other gases are released into the sea, making the bubble, for a while at least, self-sustaining.

Exploring rockpools is something of an art and an absorbing pastime. Visually a rockpool offers more in terms of exuberant colour and teeming variety of life-forms than any other habitat of its size. Contrary to widely-held belief, it is not a sinister place in which vicious biting creatures lurk in wait. Even the spines of the local sea urchin will do you no harm, particularly if gently handled, though those on the east coast can be dangerous.

The reflection from the surface tends to blur and hide the true splendour of the underwater world, as anyone who has snorkeled will confirm. A simple view-box with a glass base, or simply a pair of goggles, reveal fresh definition, colour and shape. At first glance, the life-forms are so unfamiliar and several are so well-camouflaged that it is difficult to discern their true splendid colours and sculpture; but a more careful look into crevices, beneath soft fronds of weed, and under the rocks themselves, will reveal an astonishing variety and number of creatures.

Beneath the rocks are the nocturnal and light-shy animals – including some of our loveliest molluscs – many soft-bodied creatures that must hide from the sun, and a host of tiny egg-cases and larvae which must remain hidden from the elements and from predators in order to survive. For that reason it is imperative that everything be replaced exactly where it is found, particularly stones, which must be returned facing the same way up. A delicate sea-squirt left exposed on an upturned rock cannot move away to find shelter and will soon perish.

Of the species in the Upper Balanoid, barnacles (*Balanus* species from which this zone gets its name) and limpets are most numerous. Nine species of limpet are found on this stretch of the southern coast, from the giant *Patella tabularis* to the lovely *Patella miniata* with its rosy flecked patterns. Limpets may be exposed to the air by the changing tides and have had to adapt to the stress of water-loss. This they have achieved by evolving methods of sealing their flattened shells against the rockface. Clinging like a limpet is a well-known expression, but the limpet does not use suction to do so. Instead it uses adhesion and, as anyone who has tried to prise sheets of wet glass apart will be aware, this is an extremely powerful force. In addition to adhesion, the limpets tend to return to the same place each day after their feeding expeditions over the rockface. In time, their shells grow to fit exactly the contours of their particular home site and thus create an all but watertight seal.

During the hours of darkness, limpets such as the common *Patella granularis* move slowly over the rocks, grazing algae.

Their rasping radulae leave microscopic but distinct markings on the rock and it is easy to imagine them swinging themselves slowly from side to side like miniature lawnmowers. When dawn approaches they retrace their mucous trails back to their home scars. Exactly how each one locates its scar is not clearly understood, for if the trails are erased from the rock the limpet still finds its spot and settles in precisely the same position.

The next zone is the Lower Balanoid which is exposed only at low tide. Here live several species of seaweed and mollusc. These include the *Burnupena* whelks, which are to the rocky shores what the hyaena is to the grasslands; scavengers that feed on dead or damaged animals. The whelk, *Thais dubia*, hunts here too. It is a carnivore and has the ability to drill a neat hole through the shells of limpets and other molluscs, particularly those such as mussels that are permanently attached to the rock. This whelk then inserts its specialised radula through the hole and feeds on the soft tissue of its immobile prey. Yet another of the whelks, *Argobuccinum pustulosum*, astounded scientists when they discovered it could produce concentrated sulphuric acid. They were convinced that this was used to drill through shells and yet no matter how delectable a mollusc was offered to these whelks, they refused to perform as expected and begin drilling. Subsequently it was found that the sulphuric acid was not a drilling agent at all but a means of feeding off tubeworms. The whelk thrusts its proboscis into the tube, douses the worm with sulphuric acid and shortly afterwards is able to slurp its largely predigested meal.

Scavengers must, of course, be able to locate their next meal, so whelks such as the plough snail are highly mobile, travelling freely up and down beaches in search of food. Incredibly, they are drawn by the merest whiff of a dead or dying animal and will move directly to the source. Nature's Valley, that splendid expanse of beach at the western edge of the Coastal Park, is a marvellous place to watch plough snails as they plane in on the surf and within seconds, as if drawn by a magnet, converge on a prospective meal.

Along the Tsitsikamma coast the Lower Balanoid is somewhat barren, for the whelks effectively control the number of barnacles that settle there and the limpets graze away the seaweeds. But the rockpools at this level shelter various species of which the most colourful are the anemones, not plants as so many people believe, but animals related to the

blue-bottle (Portuguese man o'war) and the jellyfish. But where the blue-bottle with its battery of angry stings is wafted freely by the ocean currents, anemones settle in one spot and depend on various methods to capture their prey. Because the rush of the waves on the Tsitsikamma coast delivers a plentiful supply of food, the anemones depend on prey being tumbled by the sea into their waiting tentacles. Where anemones in placid waters must actively hunt, shooting out microscopic stinging coils to immobilize passing creatures, those of the Tsitsikamma have only to wait for the next delivery. Instead they channel energy into anchoring themselves firmly and in developing a sturdy body to counter the effects of the waves.

Prominent residents of Lower Balanoid

rockpools are the alikreukels which have been protected in the National Park from the greedy hands of gourmands and beachcombers who savour the flesh of this giant sea snail. Elsewhere along South Africa's coast they rarely justify the description 'giant', but here they grow bigger than the size of a man's fist and every rockpool harbours several. Quite obviously in the protection of the Park they have been able to flourish, as we can but presume they would in other places were they not so assiduously collected. Thus, the Tsitsikamma Coastal National Park provides a 'witness area' which allows us to gauge the effects of man's depredations compared to places where he is not permitted to tamper.

The zone below the Balanoid is revealed only at spring low tides. Whereas

58. *The ceaseless flow and ebb of the tides veil and then reveal a varied range of habitats such as these at Blue Bay.*

elsewhere along the southern coast this zone is home of the *Patella cochlear* or 'pear limpet', so-named because of its shape, on Tsitsikamma's wave-pounded coast mussels dominate. However, in more sheltered spots *Patella cochlear* still comes into its own. No other limpet of this size anywhere in the world reaches the densities of these limpet colonies; 2 600 per square metre have been recorded. In these close quarters little else shares the Cochlear zone which, as a result, appears almost barren, sandwiched between the lush algae- and animal-filled zones above and below it.

The Cochlear zone is so packed with *Patella cochlear* that juveniles have taken to living on the backs of adults of their kind – apart from anything else to escape being eaten by their elders. One of the most obvious features of this zone, besides the sheer density of limpets, is the orderliness with which they are arranged on the rock. Close inspection reveals that each *Patella cochlear* has its own spot to which it always returns after feeding and that the home sites are evenly spaced. For the young limpets this poses a problem and they have solved it by taking to multi-storey living. Professor George Branch describes how on one occasion he found 35 individuals living one on top of the other. During high tide these juveniles come down and wander over the rockface to feed and, in the course of doing so, search out any newly-vacated home scars. If they find one and the previous tenant does not return, they waste no time in settling in. But if they are not so fortunate, they return to their multi-storey arrangement, and settle back in the identical slot they were before. How they do so is a mystery, but it evokes an image of queues of limpets waiting politely, or not so politely, until their ticket number is called and then climbing back into their regular seats.

While the limpet grazes away all algae that settle in its zone, leaving the rock apparently bare, closer examination reveals that each *Patella cochlear* is surrounded by its own garden of tiny red algae, which it not only maintains but which it protects from competition from other algae. In return, the red algae provide the limpet with its main source of food. The limpet tends its garden rather in the way that a gardener weeds his lawn and mows it continually to promote growth. In grazing the algae, the limpet actually stimulates growth to the extent that the little gardens are able to provide sufficient food for the *Patella cochlear*, thus answering the question of how they survive at such high densities. As for the

young limpets, those unable to maintain a little patch of garden on the back of a bigger limpet must struggle along, feeding on a hard white encrusting seaweed *Lithothamnium*, until such time as a spot becomes vacant and they can establish themselves as *bona fide* landowners with a garden of their own.

Brown mussels are the most striking component of Tsitsikamma's low-shore region. There they carpet the rocks in colonies so densely packed that nothing else can settle among them. Anchored by their tough byssus threads, they cope with the surge of the waves which at the same time carries particles of food to them. Rough waters favour filter-feeders and on this coast mussels have flourished to the point where little else lives below the low tide mark. But while the waves deliver their food, they also pose the largest threat. Once loosened, the mussels are helpless and the waves cut them free of the rocks, not one by one, but in great sheets. Where a vacant spot is left, other organisms settle within hours. The colony recovers quickly, however, and soon it takes over, ejecting and growing over foreign algae and molluscs until it has

closed ranks and the gap disappears.

Below the Cochlear zone is the infratidal zone which is always under water and therefore often pounded by the waves.

Besides sea urchins and an array of seaweeds, a notable resident of this zone is the red bait. Dense bands of these sea-squirts festoon the infratidal rocks, their orange-red flesh, sought after by fishermen as bait (and some gourmets as a delicacy!), hidden by a black leathery outer covering. Red bait are unprepossessing looking creatures and yet they deserve more than a moment's consideration, for in their early larval stage they have characteristics which we associate with the chordates of which man is the highest evolutionary point. The young look much like tadpoles and share with vertebrates a tail that extends beyond the anus, a tubular nerve chord down the back, a rod-like notochord, as well as gill slits. Later, when they settle on the rocks and develop into the familiar red bait they lose these characteristics, becoming immobile simple creatures. Yet they pose the tantalising possibility that should the young of a sea-squirt such as

red bait have somehow been able to reproduce itself before reaching the adult stage, we may have the missing link between our invertebrate ancestors in the primordial sea and the mammals, reptiles and man today.

These are but a few of the myriad life-forms found along the shores of the Tsitsikamma Coastal National Park. Here they are completely protected and even anglers are no longer allowed uncontrolled access.

The rocky shore is only one of the many habitats the South African shoreline offers; estuaries in particular have been proved vital to the future of our marine resources, providing shelter for juveniles of an astonishing number of species. And many marine creatures spend at least part of their life-cycles where rivers empty into the sea. Thus the proclamation of a west coast area, its ecosystems so different to those of the Tsitsikamma coast, is vital. Langebaan Coastal Marine Reserve has recently been proclaimed; but are coastal national parks enough to offer any real safeguard of future resources?

Until recently the sea was regarded as so vast and so unfathomable that nothing man could do could perceptibly change it. Today this view is recognized as false on many counts, for not only is man over-utilizing the resources of the sea, but he is polluting it. Nor are these threats which can be resolved on a national basis alone. Insecticide poured into the ocean thousands of kilometres away could well end up tainting the pilchards of our west coast – presuming that the pilchard population there recovers from our recent drastic over-exploitation. For South African marine conservation to be effective it must rest on the broader base of a global approach to managing the ocean and its coasts.

Then there is pollution closer to home. Like pollution of the oceans at large, coastal pollution is difficult to assess and to control. Nothing man could have done would have protected the Tsitsikamma coast when the tankers *Ven Oil* and *Ven Pet* spilled their cargoes of oil. Had the winds blown more of the viscous gunk ashore, much of the splendour of this coast would have been devastated.

And then there is the hinterland. The fate of estuaries proves so often the futility of protecting the sea without the same

60

environmental concern being shown on land. When a river is dammed upstream, the river mouth may gradually silt up forever, destroying the sheltered environment on which so many marine creatures depend. When a factory spews industrial waste into a nearby river and this washes down to sea, it will not be enough to have proclaimed 67 kilometres of the Tsitsikamma coast, or even 1 000 kilometres, a 'national park'.

59. *Like tiny black beads,* Littorina *snails cling to the highest of the Tsitsikamma's intertidal zones festooned with the characteristic – and hazardously slippery – seaweed,* Porphyra capensis. *Like rumpled cellophane when dried, this seaweed is eaten in Japan.*
60. *Brown mussels crowd the wave-beaten lower sea shore.* **61.** *Here in the Upper Balanoid zone* Chthamalus dentatus *is seen tightly closed at low tide. Whelks prey on barnacles such as these – which are sedentary cousins of the crabs – keeping their numbers down. Limpets graze away any seaweed sporelings that attempt to settle, leaving this zone typically barren.*

62

64

62. The docile rockcod is a colourful resident of Tsitsikamma's tidal pools.   **63.** Masters of the instant colour change that allows them to match their background – black, blue or grey as here – octopuses frequent the rockpools of the Park. A relative of the common garden snail and the limpet, the octopus is one of the most highly evolved molluscs of them all. Armed with suckered tentacles, it preys on mussels and crabs.   **64.** Many of Tsitsikamma's most wonderful marine creatures shy from direct sunlight. To find them, visitors must search beneath ledges and stones, taking care to replace everything right side up. These delicate zoanthids, anemone-like animals which stand less than a centimetre high on their sand-clad stalks, were found on the underside of a stone.

65

66

65. *The Cape sea urchin,* Parechinus, *so common in the Tsitsikamma intertidal zone, is harmless – unlike some tropical urchins which have vicious spines and may even be lethal.*

67

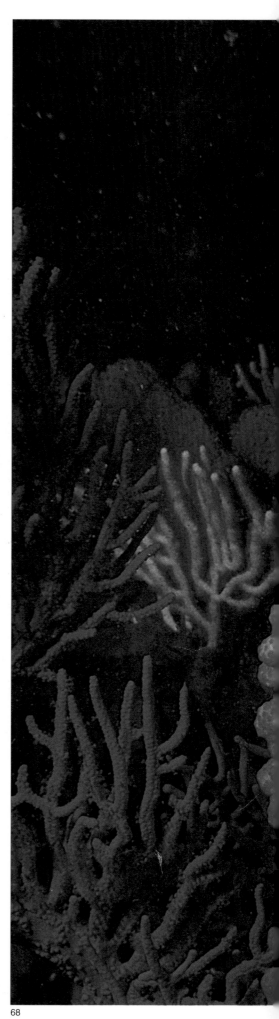

68

66. *This lovely spiny starfish is a voracious predator, particularly of the mussels so prevalent along the coast. Its main problem is in prising apart the mussel's tightly-clamped valves. To do so, the spiny starfish humps over its victim and, using its hundreds of suckered feet, exerts continuous pressure to tear the valves apart. By using its feet in relays it eventually tires out the victim. No sooner do the valves part by even a millimetre than the* starfish *exudes its stomach over the mussel – at the same time releasing enzymes that predigest the flesh before it is taken into the starfish's gut.* **67.** *As the tide retreats, sea anemones that bejewel the coast withdraw their tentacles.* **68.** *Hidden from all but the most experienced scuba divers, one of the numerous rocky reefs several hundred metres offshore blazes with a myriad creatures such as these red sea fans and purple soft corals.*

69. *For most of its length the Tsitsikamma shore is rocky, the sea seldom calm. The underwater profile drops steeply, offering little protection from breaking waves which build up over hundreds of kilometres in the Southern Ocean.*   **70.** *Kelp (black-backed) gulls scavenge along the shore.*   **71.** *The Park's famous Otter Trail takes its name from the Cape clawless otter, common along this coast but secretive in its ways. Occasionally it may be seen hunting crab, fish or octopus.*

70

71

72

73

72. Millions of years ago the elephant and the dassie (rock hyrax) shared a common ancestor and even now – despite the staggering difference in size – share characteristics such as strong third upper incisors that form tusks – small in the dassie, large in the elephant – and thick, padded soles. These adults and young are basking in a favourite haunt on the Park's rocky slopes. **73.** Over the aeons, the Storms River has cut a deep and magnificent gorge extending many kilometres upstream, which hikers on the Storms River Mouth Trail can explore. **74.** Hikers on the five-day Otter Trail enjoy a host of splendid sights such as this waterfall and pool at the sea's edge. One of the most beautiful in South Africa, this nature trail leads from the Storms River to the Groot River, crossing ravines, skirting bays and weaving through indigenous forest. Where possible it hugs the coast within sight of the pounding surf, but in places it zigzags up to the 200 metre-high plateau.

75

77

75. *These showy pincushions are familiar to many, but the Cape Floral Kingdom includes thousands of other species – no less beautiful, often small and best appreciated by a person on foot. Known broadly as fynbos (or Macchia), this complex and unique vegetation has evolved on the Cape's poor soils and is able to tolerate fire as well as hot, dry summers. Grey-green, hardy and low-growing, fynbos encompasses an array of handsome flowering plants such as the ericas, serrurias, proteas and leucadendrons. Botanists have long marvelled at the Cape's floral riches and today's visitors to the Park are equally rewarded.* **76.** *Often growing only centimetres above the high-water mark, gazanias are a feature of the Tsitsikamma shore.* **77.** *Close to Storms River Mouth, a Knysna lily splashes colour on a sun-dappled rockface beside a forest waterfall.* **78.** *Always magnificent but rough and rugged for most of its length, the Park's coastline changes mood at Nature's Valley, opening into a sweep of golden sand and a quiet lagoon that make a fitting finale to the Otter Trail.*

THE TSITSIKAMMA FOREST NATIONAL PARK

AND THE LEAVES OF THE TREE WERE FOR THE HEALING OF THE NATIONS.
REVELATIONS XXII. 2

In the damp beneath a crumbling log in the Tsitsikamma Forest, lives peripatus. You probably would not notice him, for he is a decidedly insignificant fellow; he does not snarl on the hunt nor sport massive horns; as his name – derived from *peri-* 'around' and *-patus* 'walk' – implies, he simply wanders about. But, when you realise that peripatus has been wandering for 500 million years and that he, the latest generation of his kind, looks no different from his original ancestors, he merits a second look. Not only does he exactly resemble his forebears who may well have been among the first creatures to lurch out of the primeval slime, but he also represents an intriguing link between worms and that most successful of all life-forms, insects.

The features that make this 50-millimetre creature a possible 'missing link' are invisible when you meet him in the Tsitsikamma Forest. On the outside he looks like a velvety reddish-black caterpillar moving about on stumpy clawed legs. Within, he carries the characteristics of the worms – kidneys paired at the base of each pair of legs, and a blood system – and he views his forest home through simple eyes. Yet he breathes like an insect. His soft, elastic body is permeated by trachea (breathing tubes) which carry oxygen to his system and thus he is a relative of the honey bee, the termite, the mosquito and the moth.

This unique combination may fascinate the scientist, but it leaves peripatus fatally vulnerable. Where the insect's external skeleton helps retain moisture, peripatus's soft body is easily desiccated, which means that he cannot survive outside the forest's damp, dark warmth. The other species of peripatus in southern Africa are all confined to equally damp places. This also means that for 500 million years he has been in just such an environment. Even more startling has been the discovery of his cousins in Australasia, Malaya and South America. He is so slow-moving, so tied to his humid environment, that any suggestion of his having been carried there or of having moved there himself is ludicrous.

This observation is not without significance. When Gondwana – the huge southern landmass – broke apart some 100-140 million years ago into the

79. *In the half-light of the Tsitsikamma Forest, a stream dances to the sea barely 30 metres away. 'Tsitsikamma' derives from the onomatopoeic Hottentot word for running water, and there are many streams here.*
80. *Mightiest of South Africa's trees, the Outeniqua yellowwood can reach massive proportions as in The Big Tree almost 37 metres tall and spreading into a canopy 30 metres across.*

massive fragments of land we now know as Africa, Australia, India, Antarctica and South America and these great sections drifted gradually to their present positions, they must have carried peripatus with them. How else can his presence in these distant parts be

explained? And so, the little insect-worm confirms the theory of Continental Drift.

The Tsitsikamma Forest National Park teems with plants and animals, insects and microbes, many as interesting as peripatus, many barely known. For the visitor, however, the forest evokes an emotional, rather than an intellectual response. It brings flooding back the animal instincts that helped us survive before our senses were blunted by our artifacts and our animal responses were shackled by rational thought.

Anyone who has travelled South Africa's famous Garden Route will have observed the remarkable change in vegetation along the way. As you drive east on the national road from Mossel Bay towards the Tsitsikamma, the fynbos becomes lusher and the increase in rainfall gleams from field and tree. The Southern Cape is blessed with year-round rain and regular sea mists and although the land is not particularly fertile, the moisture helps coax the maximum from it. Along the way, dense green afforestation casts shadows over broad swathes of landscape. Yet nothing prepares you for the full splendour of the Southern Cape indigenous forest. Quite suddenly the fynbos gives way and massive trees, their limbs hung with drifts of lichen, reach up and the sky recedes softly through a filigree of green.

The Tsitsikamma Forest National Park encompasses a mere 478 of the 60 500 hectares of Southern Cape indigenous forest that survive between Mossel Bay in the west and Humansdorp in the east. The Tsitsikamma Coastal National Park contains a further 530 hectares of scrub forest designated 'dry' and 'very dry' as opposed to the moist forest of the Tsitsikamma Forest National Park.

Proclaimed a national park in 1964, it protects not only the well-known forest giants – the Outeniqua yellowwood, the somewhat more modest but more numerous upright yellowwood, ironwood, stinkwood, candlewood, white pear and Cape beech – but a host of smaller trees and plants with melodious common names such as red currant, mountain saffron, tree fuchsia, rock elder and num-num. Besides the 122 woody trees and shrubs, numerous other shrubby and bulbous plants, lichens, lianas and ferns, there are 14 species of orchid, as well as carpets of mosses.

One of the most striking features of the Tsitsikamma Forest is the mix of trees; five or six specimens of the same kind often grow fairly close together but the overall impression is of variety. Furthermore, there is something eternal about these forest giants, for a tree that has taken a thousand years to reach its present size staggers the minds of mere mortals with their three-score-and-ten. And yet the reality of the forest belies this impression: besides attack by the axe in the past, the Southern Cape indigenous forest is under pressure of another kind. Over the past 2 500 years this entire area has become increasingly dry. This is not immediately apparent as you walk one of the forest paths, for what you experience will be the forest microclimate sealed in by the dense canopy of leaves overhead, by the verdant undergrowth on the sides, and by the leaves decomposing beneath your feet. At the forest edge the effects are more evident. This reality is further confirmed when you examine roads cut through the forest, for they breach the forest defences and expose the microclimate to the elements at large. In particular they allow the sunlight in, favouring weeds and other plants at the expense of tree seedlings adapted to growing in the forest gloom. Some of these plants actually inhibit the establishment of other species, while several invasive weeds compete vigorously with indigenous species.

Of course this does not happen only where roads traverse the forest; it takes place on a far larger scale along the forest margins. In its natural state a zone of shrubby plants buffers the forest from the surrounding fynbos. The importance of this zone was not fully appreciated until recently, when much of it had already been lost to fire and agriculture. Today maintenance of the buffer zone is an important aspect of park management in the Tsitsikamma and, equally, in the surrounding indigenous forest areas under Government control.

Where the buffer zone is breached, alien species take a foothold. The Australian blackwood – so well-suited to furniture-making – asserts itself, becoming a weed and a potential threat. Another weed in this area, although not of the forest itself but of the surrounding fynbos, is hakea, a well-known enemy in the Western Cape where it has taken over great stretches of land formerly cloaked by fynbos.

Hakea originated in Australia from where it was brought to South Africa as a hedging plant and to help bind areas of loose sand. Within a short time it had become a weed, for hakea is a cousin of the South African protea family and so found the new habitat entirely to its liking. Its natural biological foes had, however, been left behind in Australia.

When mechanical and chemical control of hakea proved both costly and impractical, the focus turned to the very creatures which help keep it at bay in Australia. Yet the close relationship between hakea and the Proteaceae posed a frightening question: Would an insect that destroys hakea also destroy proteas? Research provided an answer. In Australia the larvae of a certain snout beetle destroy seven out of ten fruits of the hakea and as much as 70 per cent of the remaining seed. Another insect enemy also held promise as a hakea control, for

81. *Dappled shafts of sunlight pierce the high forest canopy and play upon the tree ferns that dominate much of the forest's understorey.* **82.** *Death and decay are part of an eternal cycle played out on the forest floor. These bracket fungi help reduce dead wood to humus – the cradle in which new plants will grow.*

like the snout beetle it appeared so highly specialised that it could only survive on hakea; no other plant would do.

The links between these insects and this noxious plant merit a digression, for they reveal how research offers knowledge to those to whose care areas such as national parks are entrusted, and encourages decisions that are ecologically sound.

The female snout beetle lays her eggs on the fruit of the hakea, a plant that fruits prolifically. However, she must walk over a fruit of this particular plant or else she fails to receive some necessary stimulus and does not lay her eggs. Once the egg hatches the larva, too, is fussy. If the hakea fruit is not healthy, it will not penetrate it and move on to the next stage of its development, nor for that matter will it accept an alternative type of fruit if hakea is not available. Another intriguing aspect is that the female does not become sexually mature if she feeds only on hakea fruit. She needs to feed off hakea flowers as well before she is able to produce eggs. All this seems to confirm that the female snout beetle is no danger to other South African plant species and so can be released here without fear. But as always in the natural world there is a caveat:

evolution can confound man's plans. All it may take is for one snout beetle to survive on the fruit of, say, a bearded protea and we may find a new strain of this insect happily boring into its new host and posing a problem of its own.

Australian blackwood and roads are both potential threats to the Tsitsikamma Forest. Where does fynbos belong in this company? For many years it was believed that fynbos is a stage in the development of indigenous forest and that the two, in fact, represent different stages in the same successional trend. They do share the same habitat, and pockets of fynbos occur in the forest as well as marking its margins. To some extent the succession theory appears borne out by the fact that the tough leathery leaves of the fynbos are similar to those seen on a number of forest trees, but we now know that this relates to each having found similar answers to coping with stresses – both are unattractive to herbivores, have spiny leaves as a means of dealing with nutrient deficiency and, in the case of fynbos, have fine leaves as a strategy against drought.

More accurately, the fynbos and forest are competitors. But if fynbos appears to have the upper hand today and be on the increase, it is because climatic trends are

working against the survival of the forest. Only now we understand that the forest is a closed system, and grasp – if imperfectly – the functioning of the forest microclimate, can we begin to appreciate the effects of cutting down great swathes of trees and literally leaving gaping holes in the protective canopy, thus allowing changes in the microclimate to take place. So, too, do the effects of cutting roads through the forest take on a different perspective. In the greater part of the existing Southern Cape indigenous forest careful control is exerted, but thus far only in the Tsitsikamma Forest National Park, protected as it is by law, is there the guarantee that there will be no outside man-made disturbance of the delicate forest ecosystem.

But 478 hectares of Southern Cape indigenous forest is no guarantee of continuity without an equal degree of protection of the forest around it. As it is, this type of forest cannot increase in today's climate; it can, with wise management, be maintained at its present size; failing this it can only dwindle.

Something of the delicacy of the forest ecosystem has been revealed in recent times. The apparent abundance of growing things in the forest seems to

83. *No different to its ancestors 500 million years ago, peripatus thrives in Tsitsikamma's temperate forest, feeding off small creatures such as this centipede it has trapped with the slime it squirts at its prey.*

imply that life there is easy. Everything seems to point to immense fecundity. But the opposite is true. Two of the forest's most imposing trees, the Outeniqua yellowwood and the stinkwood, well illustrate this. It takes 50 years for such trees to become established (reach the width of a man's wrist) and these first 50 years are extremely precarious.

After being fertilised by pollen from the male tree, the female Outeniqua yellowwood, over a period of about a year, develops yellow round fruits about 20 millimetres in diameter. Within lies the seed, and it has been proved that the chances of it germinating increase considerably once the flesh has been stripped away. Here the amazing interrelationship between the forest plants and animals is revealed.

In early summer the Egyptian fruit bat dines with gusto upon the fruits of the Outeniqua yellowwood and in the course of this repast consumes most of the harvest from the tree. The bats chew away at the fleshy part and in the process drop the seed to the ground where it collects in small heaps beneath the bats' perches.

While some creatures are part of the procreative aspect of this story, others are destructive. The seed with its outer covering rich in nutrients, particularly oils and starch, is a source of food for creatures from the tiny forest dormouse to the bushpig. It must be added that, though the bushpig consumes the seed, as it scuffles about seeking these morsels it prepares the seed bed for seedling development, and the seed it fails to eat may well have a better chance of becoming established.

Should a seed survive such depredation, it will germinate close to the parent tree. And so the Outeniqua yellowwood tends to occur in clumps, often with a forest monarch at the centre.

But while creatures such as bats and bushpigs account for this pattern, birds help disperse seed over a wider area. An example is the Knysna loerie (turaco). It, too, feeds on the fruits. Later, when the seed is voided, the bird may well be some distance from the parent tree, so increasing the spread of the forest and affecting the distribution of trees.

Yet the size of the protected areas of forest we set aside as reserves and national parks is a critical factor in the continued relationship between birds and trees. The concept is simple. A bird visits a tree because of the fruit it offers. Naturally it expends energy in reaching the tree but this is offset by the meal it gains. But should the area of forest in which the bird lives become broken into small islands surrounded by fields and pastures, roads and dwellings, it may not be worth the bird's while to fly so far from one forest to another for only a limited reward in fruit and the bird ceases to feed there and disperse the seed. Furthermore once a particular remnant has shrunk below a certain size, birds may no longer be able to survive there simply because not sufficient trees are in fruit the year round to support them. Not only do the birds suffer; we now know that when this happens the future of the forest itself is in jeopardy.

In different guises, with different protagonists, this same issue arises in the setting aside of all national parks; below a certain size an ecosystem is unlikely to survive.

The stinkwood has survival strategies of its own. It has always been prized by woodcutters for its valuable timber. Yet where the trees were chopped down 100 years ago they have regrown, for the stinkwood has the ability to 'coppice' – send out new shoots – from its stump. Indeed, where the original has been felled, the root system is so well-established that the 'new' tree grows much faster than one germinated by seed. As astonishing, is the discovery that the

seeds themselves are able to coppice and send out not one but several shoots and thus increase the chances of survival.

Today, a walk along one of the forest trails reveals the stumps of trees cut by those early woodcutters. Unknowingly, they practised a conservation of sorts. Most of the trees are chopped down at close to shoulder height and the new shoots were, as a result, too high for bushbuck to nibble. Nowadays, such shoots are protected by wire mesh enclosures where trees are felled commercially (as they are in the indigenous forest outside the protection of the Tsitsikamma Forest National Park).

There are surprisingly few large animals in the park. The shy blue duiker is more often indicated by its droppings than seen; the bushpig is nocturnal and hence rarely observed. Baboon, small-spotted genet, mongoose, caracal and, occasionally, leopard may be glimpsed, while the boomslang blends so perfectly into the foliage that usually it slithers away unobserved. This lack of animals may be explained by the nature of the plants themselves, which in turn links directly with the qualities of the soil. Put another way, the forest itself is under too great a pressure to allow animals to feed upon it.

This is not a flight of fancy but a fact confirmed by research; it reveals that the Southern Cape soil in which the forest grows is extremely poor and that almost all available nutrients are, at any given time, embodied in the plants themselves. They are in perpetual use within the confine of the forest environment, and so it is termed a 'closed nutrient cycle'. Very little escapes this cycle – be it litter on the forest floor or logs decaying in the gloom. Trees such as the yellowwood waste no time in putting this to use and reabsorbing the nutrients released by decay. To do so they send up tiny roots above the surface to feed directly on the litter as it decomposes. For this reason it is vital that the mat of surface roots at the base of the trunk be protected. A wire fence restrains visitors to the 'Big Tree', not only to protect the trunk, but to protect this mat.

Given the essentially poor soil, plants can ill afford to lose shoots and leaves to herbivores and they have evolved amazing strategies to deter them. It has been discovered that their leaves contain a relatively high percentage of chemical compounds that make them unpalatable – a factor which may well account for the forest's comparatively small number of creatures that feed on plants. Taken a step further, this would account, too, for the scarcity of predators which would have hunted these creatures.

Another remarkable finding indicates that the trees of these forests appear to set seed at approximately the same time, scattering a huge quantity and providing such a glut for animals that at least a proportion of the seed has the chance to survive and germinate.

Furthermore, in many instances, the trees appear to seed at intervals of more than a year (sometimes as far apart as six years) and so place even greater strictures and stresses on the animals that feed on them. This observation has still to be fully researched, but it changes our perception of forests from that of a collection of individual trees to that of a community apparently practising 'group behaviour' of a sort we normally associate with the animal world.

The extent of the Southern Cape indigenous forest is not as great as it was even 100 years ago. Had national park status been given then, a far larger area could have been saved – both from the ravages of man and from the ravages of climatic change, for man hastened the effects of the latter. But a century ago the conservation ethic was but a gnawing doubt in those who watched the forests being felled about them.

Jan van Riebeeck spent only ten years at the Cape before he was promoted to better things in Batavia, and yet, in that short time, the forests at the foot of Table Mountain – at Rondebosch, at Newlands and at Kirstenbosch – had already disappeared. The colonists needed wood to build and furnish their homes, to make fires on which to cook, to repair their ships and, later, to fashion wagons that would carry them deep into the interior. And so they felled the forests nearby.

By the early 1700s local demand had increased dramatically and the search for timber had pressed well inland. In 1711 reports reached the Cape of immense forests in what was then known as 'Outeniqualand', prompting the arrival of the first white settlers there two years later. They set about bringing down the biggest, straightest specimens first. Yellowwood and stinkwood were the most sought after: stinkwood for its strength, fine grain and beautiful colour, coupled with characteristics that make it fairly easy to work, while yellowwood, though soft, is even-grained and a rich golden colour. Indeed, some of the finest examples of Cape furniture are a combination of these dark and light woods. Only much later was it discovered that stinkwood repels termites, and only recently has this been explained in terms of chemical substances.

Today, visitors to the Tsitsikamma Forest National Park gawk at the 'Big Tree', 36,6 metres tall, with a bole 8,5 metres in circumference and a crown broader than a tennis court. Many of these monster specimens were too large for the early woodcutters to handle. It would take a sawyer weeks to cut such a tree into planks, only to find them rejected by buyers because of the twisting and cracking common in most of these giants. As a result they tended to cut down medium-sized specimens, leaving behind the 'big trees' we see today.

After they had chopped down the medium-sized trees, the colonists pruned back the thick undergrowth and smaller trees to make space for grass on which their herds could feed. This was the established pattern of exploitation.

Some 60 years later, Governor von Plettenberg instituted a modest degree of control at Plettenberg Bay which had been developed as a harbour for the shipment of timber. If he had forebodings, they were well founded, for it had no doubt been noticed that the trees of the area are extremely slow-growing and where clearings had been made, the trees did not re-establish themselves – a fact no less true, but better understood, today.

J. F. Meeding, the man he placed in charge of the scheme, is regarded by many as South Africa's first forest warden. Within his limitations of knowledge and power he managed to achieve some balance between exploitation and conservation; indeed, it is probable that the Knysna and Tsitsikamma forests of today owe their existence to his efforts.

Meeding represented, however, a very slight force against the trends of his time. Increasingly, land was thrown open to colonists, and in 1847 the Cape Government began selling the cleared areas to farmers, thereby losing them forever to indigenous forest. And as the forest fell under the axe, fire posed another threat. This was not true of the healthy forest where there is little fuel available for a blaze – everything that is not green is rotten, and such forests only catch fire under extreme conditions. However, it was the impoverished and worked-out forests that were vulnerable and they were the prime victims. In February 1869 an immense blaze swept from Knysna eastwards to just before the Tsitsikamma forest. A second blaze, fanned by dry berg winds, surged in from the Langkloof and between them they left a blackened wasteland. The Tsitsikamma forest largely escaped the worst of these.

The 1876 discovery of gold on the Witwatersrand set the woodcutters to work with renewed vigour. The mine shafts were shored with timber and the mines needed it in vast quantities. The towns that straggled up around the mines required timber too.

By now the rape of the Southern Cape indigenous forest was so apparent that the decision by the Cape Government in 1880 to put a stop to uncontrolled deforestation met with little opposition. But it was the thinking underlying this decision itself that ultimately stayed the axe; a massive reafforestation programme was to be begun and this would, in time, provide all timber requirements for the country. The trees selected for planting were quick-

84. *The Cape dormouse, one of Tsitsikamma's small inhabitants. The forest is not heavily populated by animals, partly because much of the vegetation is rich in distasteful phenolic compounds.*

growing exotics – blue gum, blackwood and wattle from Australia, conifers from Europe and America. In 1939, having fulfilled their planned role, these exotic plantations began to reach maturity, and the indigenous forests could be appraised from a new perspective.

A result of this perspective was the proclamation of the Tsitsikamma Forest National Park in 1964. The public has a degree of access along special trails, at picnic sites and simply by driving along the Garden Route where it flanks the southern edge of the park. They also have access to other indigenous forests in the area, ones without national park status but managed with equal care.

And what of the forest peripatus? He is not big enough nor dramatic enough to merit particular attention from the conservationists. But, doubtless, as long as they look after his Tsitsikamma home, he will look after himself.

85

86

87

88

85, 86, 88. *Glorious features of Tsitsikamma's forests are the lichens and fungi. The crowns of the great yellowwoods bear drifts of old man's beard (85), and smaller vivid lichens (86,88) encrust branches and trunks. Lichens are remarkable in that they are not a single organism but an association of an alga and a fungus living together for their mutual benefit: the alga photosynthesises food for the organism, using sunlight, while the fungus furnishes its partner with moisture and shelter.* **87.** *Delicate cup fungi on a fallen forest giant, their fine hyphae penetrating deep within the timber, helping to break it down.* **89.** *Along the numerous trails leading through the dank, sweet-smelling forest, the evidence of life's cycle is everywhere. These bracket fungi and millions of unseen bacteria return fallen trees and crumbled leaves to the soil as nutrients for the seedlings that already push from the leaf litter towards the light.*

89

90. Under pressure from man's expansion, threatened by roads that open the forest canopy and expose the microclimate within, the Tsitsikamma Forest has benefited immensely from the legal protection national park status ensures. Even so the forest is more fragile than this lush and verdant view implies, for it is rooted in poor, shallow soils and functions as a 'closed nutrient cycle' in which almost all available nutrients are embodied within the plants themselves and little is allowed to remain for long on the forest floor before being absorbed into the cycle once more.   **91.** The forest chameleon spends its daylight hours stalking prey from Tsitsikamma's rich and varied insect population. In sharp contrast to its diurnal camouflage, it turns white at night when asleep.   **92.** Below the forest canopy, the bush pig rootles and snuffles through the undergrowth in search of fallen fruits and other succulent morsels the forest provides.   **93.** More often heard than seen, the Knysna turaco is a widespread resident of the forest. In flight it reveals the dramatic red wing feathers so highly prized as ornaments by members of the Swazi royal family.

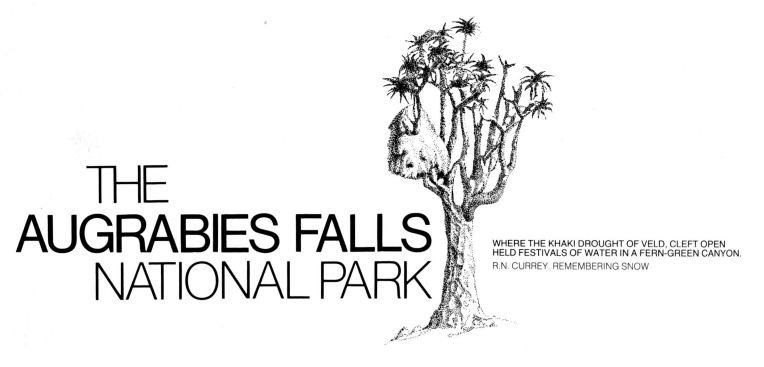

THE AUGRABIES FALLS NATIONAL PARK

WHERE THE KHAKI DROUGHT OF VELD, CLEFT OPEN
HELD FESTIVALS OF WATER IN A FERN-GREEN CANYON.
R.N. CURREY. REMEMBERING SNOW

Beyond the dusty town of Kakamas in the far north-west of South Africa the great Orange River flows smooth and slow, threading some three kilometres wide, weaving channels among low sandy islands. Then, quite abruptly, it surges over the Augrabies Falls, cascading for 25 metres and finally plunging in a magnificent vertical fall for a further 66 metres into a pool so deep it has never been plumbed – and in whose churning depths the 'water monkey' lurks.

No one has ever seen the 'water monkey'. Zoologists attribute this local myth to the huge mud barbel that certainly live above the Falls and which may well inhabit the plunge pool too. They are fiercesome, with grotesque whiskers and hideous broad black faces, and specimens two metres long have been fished from the river.

From the earliest times man has regarded the Augrabies Falls with superstitious fear. The Koranna and Bushmen who were living here when the first European saw the Falls, treated the cascade with marked respect. George Thompson, erroneously acknowledged as the first white man to gaze at the waterfall, in 1824 was led to the brink by a party of Hottentots even though 'the sight and sound of the cataract were so awful that they themselves regarded the place with awe, and ventured but seldom to visit it'.

The honour of first recorded discovery belongs to a young Swedish-born soldier Hendrik Wikar who, faced with heavy gambling debts, deserted the Dutch East India Company at the Cape in 1775. For four years he wandered along the Orange River and although his explorations ensured his welcome back at the Cape, he was not accredited with the discovery until more than a hundred years after his death when in 1916 his journal was published. It carries his original description of the Falls. He wrote: 'It seemed to me as if the whole river came tumbling down a rocky krantz twice as high as a castle.'

The impact on Thompson was equally stirring. In his excellent account *Travels and Adventures in Southern Africa* he named the falls King George's Cataract in honour of the reigning British monarch and wrote of it: 'The whole water of the river being previously confined to a bed scarcely one hundred feet in breadth, descends at once in a magnificent cascade of fully four hundred feet in height. I stood upon a cliff nearly level with the top of the fall and directly in front of it. The beams of the evening sun fell upon the cascade, and occasioned a most splendid rainbow; while the vapour mists

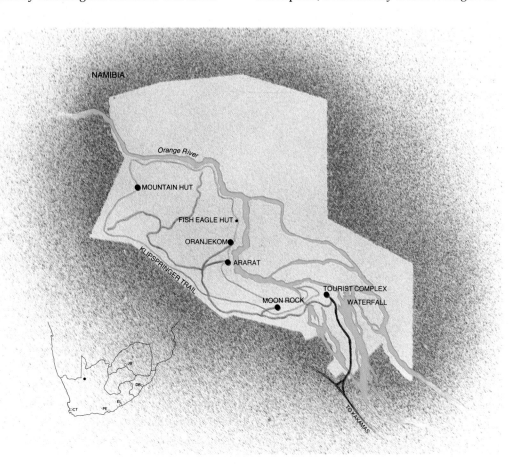

94. *Surrounded by a parched landscape almost lunar in its barrenness, the Augrabies Falls plunge into a canyon which the Orange River has carved throughout countless millennia.*

arising from the broken waters, the bright green woods which hung from the surrounding cliffs, the astounding roar of the water, and the tumultuous boiling and whirling of the stream below, striving to escape along its deep, dark and narrow path, formed altogether a combination of beauty and grandeur such as I never before witnessed. As I gazed on this stupendous scene, I felt as if in a dream.'

No one standing opposite the lip of the Falls can but share his feelings. In an arid, stark and eroded landscape beneath pale dry skies, South Africa's largest river thunders as it plunges in a spectacle of churned and frothing waters. Certainly the Koranna name *Aukoerebis* – 'place of great noise' – from which the present name Augrabies derives, is apt.

Downstream the noise decreases and the magic of the gorge grips the imagination. Here the work of water is dramatically evident for in the course of several millennia it has cut a narrow ravine 18 kilometres long running in a west-north-west direction, channelling the waters of the river on their way to the Atlantic Ocean.

The Falls was formed in the Quaternary period 1,8 million years ago when the river, as now, flowed westward across the face of the subcontinent to reach the sea. Then, some 70 million years ago, the whole subcontinent was elevated by deep and powerful geological forces, and as a result the river poured over the uplifted edge of the interior plateau. Eventually it found a nick-point – a slightly lower, perhaps even fractured, spot in the pink gneiss through which the gorge was eventually carved. As the river ate away and gradually lowered the nick-point, the might of the river was funnelled into it allowing the full force of water erosion to gouge the ravine we see today.

The original Falls was at the foot of the gorge but has worked its way back gradually to its present site. Even now erosion is at work, but the speed of the Falls' retreat is very much slowed by today's generally drier climatic conditions of the subcontinent: as recently as 20 000 years ago even dry Namaqualand received twice the rainfall it does now.

Another factor which restrains the Orange River from unleashing the sheer force of its primeval self is its upstream harnessing by immense dams.

This has meant that the flow does not change markedly year-round, whereas formerly there was a distinct season of swift, rising water matched by a season of diminished flow.

Today the regular flow of the river

95. *Wings spread for balance, a pale-winged starling departs from the coarse leaves of a quiver tree from whose flowers it has sipped life-giving nectar in this land of drought.*

ensures visitors a rewarding view of the Falls though it may not reach its former full splendour when the river would spread more than six kilometres wide and cascade down a series of lesser ravines with such beguiling names as Bride's Veil and Twin Falls.

The effects of damming the Orange have not only been aesthetic: the build-up of alluvial deposits and islands above the Falls changed with control of the water-flow and floods no longer sweep free the channels and unleash the river's full erosive powers.

The water authorities – indeed, the nation at large – have vested interest in the waters of the Orange River. It is the major supply of water for domestic and industrial use, for irrigation, and as a source of hydroelectricity for a great sector of the country – particularly the highly industrialized Highveld and the otherwise dry western reaches. It was these interests that mired the early attempts to have the Augrabies Falls National Park proclaimed; parties such as the Electricity Supply Commission and the Water Board were anxious to incorporate the Falls into the Orange River Scheme which so successfully transformed the land on either side of the river into a verdant corridor in an otherwise dry and somewhat desolate landscape.

But the ecological importance of the Augrabies area – and its relative economic uselessness – won the day, and in 1966, after government-held land outside the proposed park was exchanged for the area now incorporated, the Augrabies Falls

National Park came into being.

It encompasses some 9 415 hectares of one of South Africa's driest regions, and centres, naturally, on the Falls which is twice as high as the Niagara Falls in Canada. The gorge has a particular charm of its own. On its barren water-cut walls rock-splitting wild Namaqua figs add their forces to that of the water to change the face of the landscape, their white powerful roots clinging to the surface and prising apart fissures. Black storks nest on tiny ledges and the distinctive call of the fish eagle echoes from the sheer pink walls. All around pale-winged starlings whistle and call, and the startling gunshot clap of the rock pigeon's wings rings out as it takes off in flight over the void.

Visitors to the Park make the Falls their first point of call, observing it from several vantage-points protected by a fence (for 16 lives have been lost by the foolhardy trying to peer over the edge and losing their balance). But there is far more to this Park than just the Falls and its gorge, magnificent as they are. Several roads direct the visitor to raw and rugged landscapes such as that around Ararat, Moon Rock and Echo Corner. Another more intimate view is gained from the specially laid-out footpaths or by setting out on the three-day Klipspringer Trail.

Spring and autumn are the best times to take this trail – indeed, to visit the Park – for this part of the country is not only dry, but experiences trying climatic extremes: between night and day the temperature may fluctuate by 20°. The bitter chill of winter nights is matched by warm days, but the cool nights of summer contrast uncomfortably with daytime temperatures that can soar above 40 °C. In summer the plague of swarming 'muggies' (midges) can be severe and the already forbidding aspect of the Park takes on a certain merciless quality before which most visitors would quail.

The Orange River bisects the Park into almost equal halves, of which the southern – the area open to exploration – is the most rewarding. Between the gorge and the edge of the Bushman peneplain (the ancient platform or foundation of the subcontinent) is a transitional area of great interest and unyielding beauty. Here are giant low-lying dome-shaped rocks, of which the most famous is Moon Rock. The immense parabolas of these outcrops are the result of exfoliation – the flaking off of great slabs of rock much as the skin of an onion lifts away layer by layer. Temperature fluctuations cause this particular kind of erosion: while the interior temperature of the rock remains fairly constant, the outer surface expands

in the heat and contracts with the cold, eventually becoming loosened and 'peeling' away, leaving it typically rounded. Beneath the lifting slabs shelter several intriguing creatures including the red-tailed rock lizard and the rare, vibrantly-coloured red-banded frog. Like all frogs, it breeds in water, an element in short supply where it lives. However, even the lightest shower of rain tends to collect in depressions in the rocks and attracts insects which provide a ready supply of food for the frog's tadpoles.

Dassies, or rock hyraxes, scurry on the hillsides, keeping a wary eye for the many birds of prey that patrol these skies. Occasionally a klipspringer may appear briefly in silhouette and then, with a delicate arabesque, disappear from view. Snakes and lizards bask in the sun, drawing on the heat to warm their reptilian blood. But the landscape tends to be silent and still during the day, for most of the insects, reptiles, rodents and other small mammals avoid the stresses of its heat. Many hide in crevices, in burrows or beneath the sand, and emerge only at night. The bark of a male baboon attracts the visitor to a kokerboom in bloom, the baboons feasting on the nectar-rich flowers, licking their lips with relish at the sweet bounty. Their enemy the leopard is at rest in some shady retreat in the hills, but with the sun up in the sky the danger is small; at night the baboons will keep a cautious watch.

Not all the hills are exfoliated, nor are they all the prevailing pink-orange so characteristic of this Park. More than half the Park is of rock and most of the rock is pink gneiss, much like granite in mineral content and chemical composition. In striking contrast to the gneiss, however, are the black quartz-rich granulite outcrops making low hills, often scattered with chunks of exquisite rose quartz. Worn from the hills are poor pale sands, thin and infertile. In the west and south-west the shallow soils are topped by a layer known as Schaumbaden: the outer crust is a few millimetres thick below which lies a layer with a foam-like structure which water cannot penetrate, thus leading to quick run-off and sheet floods.

Drainage lines are an important focus of plant-life in this Park, where shortage of moisture is so critical to survival. They are often lined with thickets and open

96. *Rearing from the bleak, dark-rock crumble of the Augrabies Falls National Park, the quiver trees seem the ancients of survival. Their common name derives from the Bushman (San) practice of using the bark to make quivers.*

shrubland. Under low trees other plant species thrive because there may be as much as a six-fold increase in the water percolating into the soil below them. Furthermore, beneath the canopy of the tree the extremes of temperature, evaporation and humidity are eased to some degree, making for a slightly kinder environment than that which seres and scorches beyond its shelter.

Above the Falls and along the river the vegetation is spared some of these stresses experienced, and tolerated, by the vegetation elsewhere in the Park. The western and northern sandy areas as well as those south of the river tend to support grasses and a variety of shrubs with succulent or leathery leaves and stems. Here grow trees such as the smelly shepherd's tree, the driedoring found also in the Kalahari, the leafless cadaba and the sjambokbossie.

The most impressive of the plants in the Augrabies Falls National Park is the kokerboom or quiver tree, handsome with its thickset massive trunks with flaking pearly-grey bark. Many fine specimens grow proud and tall on the rocky slopes, their branches sometimes hung with a heavy social weaver nest. Another characteristic tree is the scraggy-looking wolftoon, while the spekbos, shepherd's

tree, Nama resin tree, Bushman's tea and melkbos are but a few of the species common to the area.

Like arid areas the world over, the Augrabies Falls National Park is fragile, its ecological balance easily disturbed and difficult to manipulate. Even the reintroduction of larger mammals such as the red hartebeest must be done with caution. To date away from the river itself the Park has been little affected by man and perhaps more than any other of South Africa's dry regions it evokes a deep and mystical response in the face of one of nature's most impressive handiworks.

97

98

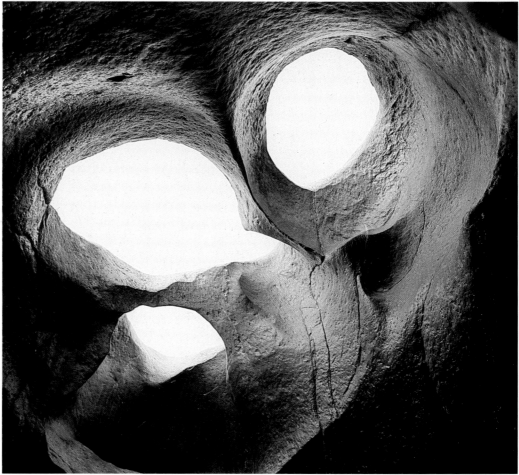
99

100

97. The spume of the roaring waters creates an environment in which clouds of midges breed and swarm, providing a constant supply of food for Platysaurus capensis, a multi-coloured lizard which scampers across the canyon's walls defying gravity as it hunts its prey. **98.** Infrequent rains and a few drainage lines bring a touch of green to the drab browns and greys of the Park – and provide food for the springbok which move across the barren landscape. **99.** Over countless aeons the force of the Falls has driven loose boulders to gouge enormous potholes, at the bottom of which a wealth of alluvial diamonds reputedly lies. **100.** Silhouetted against a typical Park setting of bare rock and equally bare thorn bush, a male and female klipspringer pause cautiously to view their surroundings before bounding off on hooves delicately adapted to their rocky habitat. **101.** In full spate the waters of the Orange are an awesome sight, leading the early traveller George Thompson to enthuse: 'As I gazed on this stupendous scene, I felt as if in a dream.' Here the smaller Bridal Falls join the river just below the main cataract on the right.

101

102

103

102-104. *Rays from the setting sun pick out the hillocks of rock that are a feature of the arid landscape of the Augrabies Falls National Park (104). Rain seldom falls here and drought will drive even the sociable weavers from their communal homes high in the crown of a quiver tree (103). Yet a few creatures cling to the sere habitat and a rosy-faced lovebird (102) has opted for squatters' rights in the deserted weavers' colony.*

104

THE BONTEBOK
NATIONAL PARK

WHAT IS WILD IS, BY DEFINITION, UNMANAGEABLE.
I.S.C. PARKER. CONSERVATION, REALISM AND THE FUTURE

Like a small but beautiful jewel, the Bontebok National Park rests in its splendid setting below the Langeberg range and only six kilometres from the historic and lovely town of Swellendam. Covering a mere 28 square kilometres of south coast renosterveld and traversed by the broad and sometimes strong-flowing Breede River, it illustrates another facet of the many purposes and principles of national parks: it has been set aside expressly to protect and guarantee the survival of the formerly endangered bontebok.

The bontebok avoided only by a hair's breadth the plunge into the abyss of extinction and even now must rank as the rarest antelope in southern Africa. Like the luckless blue buck – a handsome blue-grey animal related to the stately roan and sable antelopes farther north – the bontebok was pre-adapted as a candidate for extinction. Both had extremely restricted ranges which happened to coincide with the areas coveted by the earliest white settlers. The blue buck was wantonly eliminated; the bontebok survived.

From the historical literature it seems that at the time of first European settlement the bontebok's range was a mere strip 270 kilometres long and 56 kilometres wide, bounded by the Langeberg in the north, by the Bot River in the west and ending at Mossel Bay in the east. Within so limited an area it is unlikely that the bontebok was ever numerous. Furthermore, early estimates are distorted by the fact that many writers

could not, or did not, distinguish between the bontebok *(Damaliscus dorcas dorcas)* with its stronger colour ('*bont*' means many-coloured) and shining white rump, and its close relative the blesbok *(Damaliscus dorcas phillipsi)* which is pale brown on the rump. The blesbok is also distinguished by a broken white facial blaze from which its name derives. The blesbok's range is generally accepted to be the plains of what are now the Free State, northern Cape and the southern Transvaal.

Both antelope are closely related – indeed, they are subspecies of the same genus – and will readily interbreed, as has happened on several farms where they have been allowed to mix. Many hybrids have been killed and farmers, urged to destroy more, are being made increasingly aware of how undesirable interbreeding

is; a pure and separate gene pool for each is a far better blueprint for survival of the species as a whole.

The bontebok was threatened by the same indiscriminate hunting and enclosure of the land for farms as the blue buck, but though by 1930 its numbers were severely reduced, a few remaining herds still eked a precarious existence on the relatively unpalatable vegetation.

Fortunately a handful of enlightened landowners in the Bredasdorp area – notably the Van Breda, Van der Byl and Albertyn families – were aware of the bontebok's decline and actively protected as many as they could on their own land. However, the situation was desperate, and with the fate of the blue buck a stark reminder, proclamation of the Bontebok National Park in 1931 was timeous – and vital.

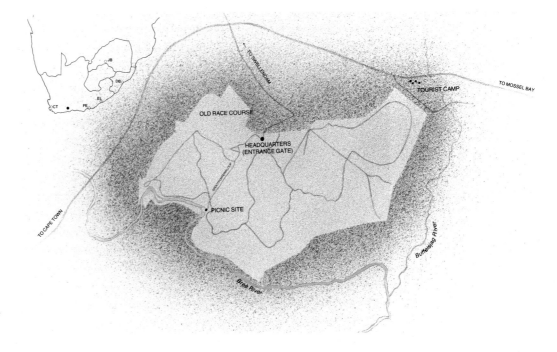

105. *A white blaze endows this bontebok with a certain dignity. Until recently, this species of antelope faced extinction.*

106. *Spotlighted by a burst of sun, a group of bontebok grazes on the stretch of coastal veld set aside as their second sanctuary when the first Park was found, among other things, to be deficient in the trace minerals essential to their health.*

The area set aside was not that occupied by the present Park: the Bontebok National Park has the dubious distinction of being the only national park in South Africa to have changed site. The original 722 hectares situated 27 kilometres from Bredasdorp were stocked with a paltry 22 bontebok. The animal's fortunes were at their lowest ebb, for not only were their numbers few, but the land set aside for them was clearly unsuitable. After an initial spurt this nucleus bred poorly, was heavily infested by parasites and suffered from 'swayback' – a general weakness caused by a deficiency of copper in the diet. Obviously, if the Park was to maintain a vigorous and genetically pure bontebok population, something had to be done.

A solution was long coming and only in 1960 were the 84 animals in the old Park taken by truck to the newly-acquired site near Swellendam. In 1961 the Bontebok National Park was proclaimed for the second time.

In their new home the bontebok have thrived, and in October 1981 there were some 320, with many more translocated to zoos around the world and others distributed on nearby farms so that numbers could be kept reasonably constant within the Park (which has a limited carrying capacity) and to establish alternative breeding nuclei.

The Bontebok National Park has posed several interesting problems since its inception: firstly in its move, but more lately in various other ways. Without doubt the Park is too small to maintain 500 bontebok – according to current scientific theory, the critical number that guarantees the future of a species. At best it can serve as a 'stud farm', maintaining a healthy breeding stock from which herds can be built up elsewhere, improving the antelopes' chances of survival. Where these nuclei should be developed is a thorny issue. One alternative is zoos; a somewhat tame and ignominious fate for any wild animal. Another option has been to give local farmers pairs of bontebok to breed. This was largely unsuccessful: the bontebok is a social animal and requires a social group of a certain composition of sexes and ages to thrive. While farmers were prepared to set aside sufficient land for two animals, they were not prepared to settle several on a tract sufficiently large to accommodate the male bontebok's territorial behaviour. An answer lies in increasing the size of the Park. But it is surrounded by excellent farmland and farmers are unwilling to part with holdings which can yield profits in wheat and wine that no bontebok could match. Furthermore, additional land is not an instant solution as it would take many decades to rehabilitate and resemble its former state.

The current situation is a compromise and, given the small area of the Park, the specialised habitat requirements of the bontebok and the value of the surrounding farmland, it is unlikely that a better alternative will emerge.

Contrary to popular belief, single species conservation does not – even with the 'magic' number of 500 of a species – ensure a strong and genetically diverse population. The main reason for this is the absence of predators. We tend to view predators in a somewhat negative light, as the consumers of animals. In the case of the bontebok which is already diminished in number and habitat, lack of predators seems a relief. Yet, paradoxically, predators have a honing effect that improves and strengthens their prey. For example, the social composition of the herd, the bontebok's cryptic patterning and splendid colours may all have evolved in response to the threat of predation. Without this threat, and without the natural culling by predators of the weak and unfit, the bontebok is deprived of one of the most telling imperatives of survival that shaped it thus far.

The bontebok we see today is the triumphant result of the cosmic dice-game of evolution. Over thousands of years it has managed at once to breed, to survive, and to avoid being killed. In the past it would have had, as well as man, various enemies such as lion, leopard and hyaena. Today in the protected environment of the Bontebok National Park we have conserved half the equation: we have rejected the predator but maintained the prey. And, not unexpectedly, the bontebok we are conserving may become but a pale representative of its kind.

The Park is also too small to be 'rounded out' or stocked with a variety of species, including those that may have occurred here in times past. At one stage the Parks Board introduced animals as diverse as the Cape buffalo and ostrich,

red hartebeest, eland and springbok. Without doubt, however, under the combined pressure of the bontebok and the host of introduced species, the vegetation continued to deteriorate and the principal objective of the Park, the bontebok, was potentially threatened. Acknowledging the situation, the Park managers then removed from the area all the introduced species but for the bontebok. Besides the bontebok, the steenbok, Cape grysbok, duiker and the grey rhebuck were allowed to remain, firstly because they were already there, are not particularly numerous and also because they do not appear to severely damage the plant-life. The graceful but aggressive grey rhebuck is particularly conserved in the Park and can be seen in parties of up to 30.

But if the variety of animals on display below the Langeberg is strictly limited, it must compete with a far more varied and perhaps more beautiful star: one whose rating and appeal have risen in recent years and whose future survival is equally important – the fynbos. This is the vegetation of the Cape Floral Kingdom, sixth and smallest of the world's floristic kingdoms and at the same time its most diverse and wonderful. It is unique to the southern parts of the Cape Province, and generally restricted to the winter-rainfall regions although a marginal form is splendidly represented in the Bontebok National Park, where half of the rain falls in summer and spring.

Of all the floristic kingdoms, this is the one most under pressure: and this pressure is not new although it is increasing at an alarming rate. Week by week populations of species are

eradicated by farmers, by builders, by roads and man's careless impact on his environment.

The fynbos and grasses now incorporated in the Bontebok National Park were not exempt from these pressures. The land has been planted with foreign plants, has been invaded by alien vegetation, and burnt to encourage grazing for domestic herds; the western half of the present Park has even been used as a race-course.

And so the focus of the Park's managers has moved from a total preoccupation with the bontebok, the grey rhebuck and other animals to a concern for the vegetation. Inherent in this is an element of conflict, for the bontebok requires sweet grass if it is to thrive; yet the fate of the fynbos is also paramount. If the Park were larger, a compromise might be reached more easily. But the issue is further compounded by the fact that because no one can be certain of the pristine state of the area's vegetation, it is also difficult to ascertain a 'natural' percentage of grass to fynbos. To allow the vegetation to evolve naturally may result in less grass (particularly the nutritious *rooigras* which the bontebok favours) although prior to European settlement there seems to have been more. As it is, the present composition of the Park is low shrub, dominated in parts by the tangled grey-green renosterbos (so named for the early explorers' finding rhinoceros hiding in it). Along the Breede River grow fine *Acacia karroo*, Breede River yellowwood, wild olive and milkwood which provide shade and cover; while the ridges and gravel terraces are cloaked in the attractive yellow-greens of

leucadendrons. On the slopes grow other members of the rare and lovely Proteaceae, including *Protea repens*, floral emblem on the South African coat-of-arms and a source of sugar for the early settlers, who boiled down the copious nectar these plants produce.

Today sectors of the Park are burnt each year, to a specific rotational plan. Fynbos is adapted to periodic burning – indeed, some species require to be burnt before they will come into flower. However, it may well be that for the Park to succeed as a bontebok stud farm, the requirements of the fynbos must become secondary.

It has been shown, for example, that small rodents (such as the Namaqua rock mouse, Verreaux mouse and Cape striped mouse) and birds are pollinators of the fynbos and are important factors in its successful survival. But the sugarbirds and sunbirds will not travel far to feed on – and at the same time pollinate – species of fynbos if the plants occur in such small groups that it is not worthwhile (in terms of energy expended to food gained in reward) for the bird to make the trip. As in the southern Cape indigenous coastal forest, the 'island' syndrome is a threat: in the Bontebok National Park, the destruction of fynbos on the surrounding land may leave the Park as a patch too small and too distant from another for the pollinators to visit, ensuring the eventual demise of this special component of the Park's vegetation.

But while these problems trouble the Park managers, visitors are still entranced by the many and various attributes of this national park: its natural loveliness, its population of graceful antelope and the splendour of its vulnerable plant-life.

107

108

107-111. *Tissues of mist lingering above the Breede River (108) form a hesitant veil to the Langeberg range in whose shadow lies the Bontebok National Park, where the red spears of Aloe ferox stand lonely sentry duty. The river's banks are the habitat of a wide range of creatures from the tiny and delicately-patterned striped stream frog (107) to the subtly camouflaged Cape reed snake (111), another waterside denizen. The aloe sentinels make a convenient resting spot and vantage-point for the Cape weaver (109) while between their leaves, or any other convenient vegetation, the orb web spider (110) will spin its food-trap.*

109

110

111

113

112

114

115 ▶

112-115. *Stately, almost arrogant, a blue crane (112) stalks across the spring-carpeted veld, ever alert for the insects or small rodents which form much of its prey. In spring the fynbos of the Park shows its true glory: the bright hues of the mesembryanthemums (113) patch the veld in gleaming colour; the delicate flower heads and bracts of leucadendron (114) nod to the wind; and the flowers of plants such as this Leucospermum reflexum (115) create a larder for countless sugarbirds. But the fynbos and the bontebok share the Park uneasily, the conservation needs of each posing a threat to the other.*

116. *While other members of the family group graze unconcernedly, a female bontebok suckles her lamb. In this safe and suitable environment the bontebok herds are thriving – so much so that their numbers already put pressure on their living space, though their population has not yet reached the magic 500 which geneticists believe essential for survival of a species. Now conservationists must decide whether to disperse the herds or try to enlarge the Park.*

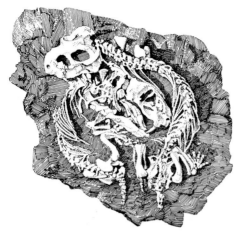

THE **KAROO** NATIONAL PARK

THE KOPPIES THRUSTING TOWARDS HEAVEN
THEIR BLUNT GRUDGING PRAYERS, THE SMALL WHIRLWINDS
AS IF THE LAND WERE TRYING TO SPELL OUT POEMS.
PERSEUS ADAMS. THE WOMAN AND THE ALOE

The grand horizontal sweep of the Karoo landscape, hard blue skies above, koppies blocked low on a bleached horizon: impressions of a strange beauty, memories of dryness and monotony. The ordinary traveller crossing the Great Karoo today may well ask what there is to conserve here. Why have a national park in the Karoo?

Yet this arid area, so vast it accounts for almost a quarter of South Africa, is in some respects one of its most fascinating

and its most powerful. For many South Africans it is their spiritual home, its grand and empty landscapes evoke the Afrikaner's first forging of emotional links with the Karoo. If ever the Trekkers' courage and resolve to beat a pioneer path to the interior was tested, it must have been from 1836 to 1838 when they first trundled their clumsy ox-wagons over the Swartberg onto these daunting plains – and decided to press on.

Today the white-washed farmsteads,

dwarfed and lonely, pluck a responsive chord from the national psyche. Beside many of these homes willows relieve the Karoo's drabness, for the Trekkers, and those who chose to settle here, often planted these trees beside perennial springs, perhaps as a beacon to travellers across the plains. The willows remain but many of the farmsteads are deserted and crumbling, and most of the springs have dried up. Today's Karoo farmer draws water for his family and for his sheep and goats from a reservoir of corrugated iron filled by the deep suck of a Karoo windpump, its sails moving slowly round and round, scarcely disturbing the prevailing stillness.

And if the Karoo stirs a fierce sense of freedom and endurance, it also nudges the curiosity – and richly rewards those who persevere and learn its secrets. The vegetation, for instance, so squat and drab when seen from the window of a moving motor-car, needs to be examined close up to be appreciated. Here you will discover an array of plants wonderfully adapted to the Karoo's unforgiving climate. Botanists term much of it 'xerophytic', meaning adapted to lose very little water to the air and able to survive in extremely dry conditions. Knowing this, we learn to appraise anew the fleshy leaves and moisture-filled, rounded stems of plants such as the mesembryanthemums and stapelias; the reduced and all but invisible leaves of the bushman's candle; the

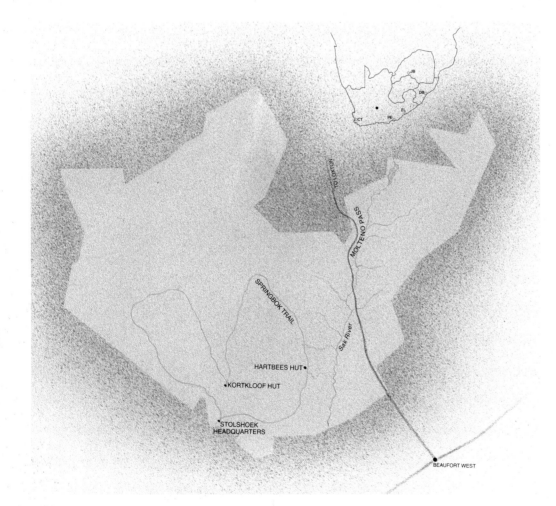

---

117. Boulders, flaked by the extremes of temperature which the Karoo experiences, seem to dwarf the crimson-flowered wild broom whose copious nectar attracts bees and sunbirds. Also known as 'swart storm', Hottentots used its bark to concoct a purgative.

118. *Pairs of angry white thorns dissuade all but the most persistent herbivores from browsing the Karoo acacia whose yellow pompom florets bring startling blotches of colour to the dark palette of the landscape.*

thickened bulbous roots of the stumpy melkboom; and the tiny, leathery leaves of the crassulas.

Many of the plants have a waxy surface to their leaves, as a means of cutting down transpiration (the loss of moisture to the air), coupled with the ability to plug or close the stomata or tiny surface pores through which plants exchange carbon dioxide and oxygen with the air. Just as these are means to conserve moisture, so a number of plants have also evolved means to protect themselves, especially from grazing or browsing animals. As a defence some are unpalatable, but others are in fact very nutritious and arm themselves with thorns, barbs and razor-edged leaves.

The grasses cope in a different way: where other plants endure, several have life-cycles which avoid the worst conditions. Termed 'ephemeral', they grow vigorously when rain falls, set seed within days and then wither, their seeds lying dormant until conditions favour them once more.

All the Karoo plants have life-cycles geared to the rare but often violent thunderstorms. Then pure pink and orange blooms splash the dour Karoo palette; bushes carry bright daisies scattered and brilliant like the stars of a Karoo night; the grasses shimmer with seed and the landscape rejoices – for it may be a long time before rain falls again.

Another not immediately apparent facet of the Karoo is its wealth of fossils which have contributed significantly to our understanding not only of prehistory, but of the world as it is today. In the words of Dr G. Thom (1789-1842), one of South Africa's earliest palaeontologists, it lays 'open the evidence of an ancient ocean . . . which has left in letters of stone the traces of a former world, so that the persevering observer, with an ordinary share of intelligence, may read and understand'. Perhaps he over-simplifies the case for the study of fossils is a highly specialised science, but the value of these ancient remains locked in the Karoo sandstone and shales is immeasurable.

And then there are the Karoo skies, far from city lights and smog. Beyond their clarity, scientists probe the night from South Africa's most powerful telescope at the observatory at Sutherland (also known to experience the lowest winter temperatures in the country). Even to the most casual gazer, the magic of a star-strewn Karoo night is unforgettable.

It is something of a similar magic that the Karoo National Park is intended to conserve. Proclaimed in September 1979, it encompasses some 20 000 richly varied hectares, representative of the Karoo's unique interplay of climate, topography and geology. The Park's profile includes the high-lying area above the steep Nuweveld escarpment; the escarpment itself with its deep-cut ravines; and the slopes reaching down to the plains in the south and south-east dotted with characteristic Karoo koppies.

This landscape just north of the agricultural town of Beaufort West supports many different communities of plant-life all equipped to endure a climate whose daily temperatures fluctuate spectacularly, droughts that drag on for years, exhausting summer heat and bitterly cold winters that shroud the escarpment in snow and lay a brittle cast of frost on the plains. And when rains do fall, thunderclouds release torrents that scour the land.

The National Parks Act of 1962 recognised the need to conserve a representative – and viable – sector of each of South Africa's many ecosystems. One of these is clearly the immense inland basin of the Great Karoo, lying south of the Nuweveld escarpment and stretching far south to the Swartberg. Yet ten years later when Dr R. Knobel, then Director of National Parks, made the establishment of a Karoo national park a top priority, the Karoo of old had already long since disappeared.

As interesting as the area of the present

Park is, it is but a poor reflection of the Great Karoo the Trekkers found. The climate was no kinder; long before the Trekkers came here the Hottentots (Khoisan) had named it *kuru*, meaning harsh or dry. But it had another aspect difficult to imagine today. It was cloaked in grasses and dwarf shrubs that flowed across these same plains where today ubiquitous grey-green bushes shroud to the horizon. On these grasses and many succulents, grazed great herds: springbok, wildebeest, ostrich and the ill-fated quagga. In the mountains there were leopard, and on the plains lion prowled. The springbok population alone must have been phenomenal. Writings of the time record their periodic mass movements to new pastures and they astound the mind. In 1849 Sir John Fraser observed a herd of 'trek-bokke' so vast that it took three days to pass through Beaufort West. A year later, S.C. Cronwright-Schreiner, husband of the famous South African author Olive Schreiner, observed one of the last springbok treks and estimated the mass of animals to be 220 kilometres long and 21 kilometres wide.

These estimates may be generous, but the existence of such huge herds is without doubt – and without recent parallel. Indeed, the Karoo we see today is largely bare of wild animals bigger than a duiker.

The predators were destroyed by the herders who, with baited traps and bullets, ensured the demise of the lion, the hyaenas, wild dog and even the magnificent lammergeyer, a bearded vulture-like eagle accused of killing lambs and hence summarily despatched on sight. A few leopards still stalk in the kloofs, but outside the Park their continued existence is precarious.

The antelope met a similar fate. They were shot out in hunting orgies, by travellers, armies and by black tribesmen in need of food. Later they were slaughtered by the authorities in an attempt to halt the spread of stock diseases such as rinderpest. It is a chilling thought that these antelope could not survive on the present impoverished Karoo in anything resembling their former numbers – the grasses and sweet succulents, the perennial springs and waterholes are too depleted.

Cattle were the first significant livestock imports to the Great Karoo, along with the fat-tail sheep. Both trekboer and the early tribesmen were herders, and the low-lying basin of the Great Karoo must have offered great promise – endless grasses, rich succulents, springs edged by trees suitable for firewood and for building. These early attributes are hard to imagine now: the few remaining trees are cut down to fuel the cooking fires of people too poor to afford anything else; the springs have dwindled, their moisture seeping through bared and eroded earth. As for the shimmering grasses and the host of other highly nutritious plants, they are gone, just as the antelope that grazed them are gone. In their place are sheep and goats.

These two animals are blamed for the deterioration of the Karoo, for the stripping of the earth so that water run-off (and loss of top-soil) have become such a threat. Yet the most damning factor has been the fence – coupled with human greed. It has encouraged farmers to restrict great herds of domesticated animals on land which is not only unable to carry the numbers but cannot bear the brunt of constant grazing. When the grasses could no longer sustain cattle, merino sheep were penned in the Karoo. The wool market, though a bonanza for the Karoo farmer, tempted many of them to risk the long-term destruction of their land for the short-term profits the wool mills of the world have offered. And they have paid the price: even sheep struggle to find enough grazing in today's Karoo and that unfussy, opportunist feeder the goat has been introduced to take the last few bites.

Despite this, the Karoo of today remains a place of fascination and to look down from the great Nuweveld escarpment in the Karoo National Park it is easy to imagine that the wide vistas have not changed in hundreds of years. Yet even as the Karoo's huge herds of springbok were declining before the advance of European settlement and technology, a part of that same technology was instrumental in uncovering a whole new world of Karoo creatures, frozen into time itself and a very part of the rocks and koppies of the Karoo landscape.

In 1838 Andrew Geddes Bain, who had already earned a wide reputation as an explorer and road builder, was working on the construction of a road, north of Grahamstown, which passed through sedimentary rock of what is today known as the Karoo Formation or Supergroup. He had already developed an interest in geology and was well aware of the possibility of uncovering the fossilised remains of animals and plants in the course of his road-building. Eventually, in the vicinity of Fort Beaufort he found the fossil skull of a strange reptile with only two teeth in the upper jaw – an animal he immediately named the 'Bidental' ('two teeth'), today known as the *Dicynodon*.

His enthusiasm sharpened, he soon amassed a considerable collection of fossils from the Karoo rocks and was eventually obliged to hire a room in Grahamstown to store them.

Bain was in the unenviable position of discovering what he realised were items of immense scientific interest, yet having no-one in the country who could properly study and describe them. In frustration in 1844 he sent his fossils to England but not before holding a special exhibition at Grahamstown where they aroused much interest.

In England Bain's fossils were recognised for what they were: remains of an entirely new and hitherto unknown order of animals not represented by any living descendants.

With the hindsight of nearly one and a half centuries, we can view Bain's work in perspective and appreciate the magnitude of his discoveries. His 'Bidental' and other fossils represent an extinct group of reptiles today known as the mammal-like reptiles or Therapsida. Mammal-like reptiles were, as their name suggests, transitional between true cold-blooded reptiles characterised by scaly skins and an egg-laying mode of reproduction, and the first mammals characterised in turn by warm blood, furry skins, and milk glands with which to suckle live-born young. True mammals arose some 200 million years ago; Bain's mammal-like reptiles which show a combination of reptile and mammal characteristics were ancestral to true mammals and lived 230 million years ago.

For the last half of the nineteenth century descriptions of the unexpected wealth of Karoo fossils kept the pens of the world's leading palaeontologists busy. Gradually the extent of the fossil-bearing rocks was discovered and so were their ages. And from this study the scenario of the formation of the Karoo rocks and the origin of the numerous fossils have been established.

Some 250 million years ago, during the Permian period of geological time, southern Africa was emerging from the grip of a great ice age. As the widespread ice sheets retreated, the interior of the subcontinent was revealed as a low-lying region into which numerous rivers and streams flowed. This immense natural depression, which today is referred to as the Karoo Basin, was a land of marshes, pools and low stretches of dry land subjected to regular flooding. At first the landscape was dominated by plant-life, strange by comparison with today's in that there were neither flowering plants nor grasses. These plant forms were yet to evolve many million years later, in the Cretaceous period.

Some time after the luxuriant plant-life had become firmly established in the Karoo marshes, the first land-dwelling reptiles made their appearance – following what appears to have been a long trek from distant central Europe and Russia, where similar but slightly older fossils are known. Many of these animals were large and ungainly, but among them lived smaller more agile forms. Some were herbivores, others were carnivores

119. *The vividly-contrasting red and black colours of this pyrgomorphid grasshopper warn potential predators that it would provide a highly distasteful meal. Despite the Park's apparent barrenness and frequent lack of rain, it is home to a myriad insects, reptiles and other small animals.*

and studies of even this, one of the very earliest southern African ecosystems, reveals the finely balanced ecological relationships between the various species of animal and plant.

How were the remains of these prehistoric Karoo reptiles preserved as fossils and how did the present-day Karoo landscape, which resembles anything but a marsh, evolve? Research has shown that sand, mud and clay carried into the ancient low-lying Karoo Basin consolidated, and these layers led to the formation of the sandstones and shales so characteristic of the Karoo.

Included in them are the petrified remains of the animals that died, and whose carcasses became buried in the fine silt. Protected within their muddy tombs from the destructive effects of sun, wind and water, as the mud and clay consolidated, these bones underwent gradual chemical changes ensuring their preservation as fossils during the coming millennia. Deep in the earth, covered by later layers of sediment, they would lie undisturbed while continents broke apart, mountain chains formed and eroded away, and ice ages came and went.

Fossilisation has been in progress since the earliest forms of life appeared on earth – and it continues today on the bottom of sea and lake. But the Karoo succession of rocks is unusual and important in that the sedimentation process continued almost uninterrupted for nearly 50 million years, covering the later part of the Permian and the entire Triassic periods. It resulted in alternating shale and sandstone layers thousands of metres thick and unique in the world: nowhere else on earth during these periods were sediments of this kind accumulating on anything near a comparable scale.

The fossils in these rocks are also largely unique; although similar and even the same animals and plants must have occurred elsewhere, their remains were not often preserved, were weathered away long ago when the forces of erosion sculpted the earth's surface, or have not yet been uncovered.

The reason for the modern Karoo landscape's total lack of resemblance to this prehistoric scenario is evident from the uppermost, or most recent, rocks. They reveal that about 200 million years ago (in the late part of the Triassic period), dry even semi-desert conditions prevailed. Also evident from the very uppermost of the Karoo's sedimentary rocks is the onset of volcanic activity with frequent lava flows that hardened into basalt. The Karoo Basin was eventually overwhelmed by these massive lava flows that spread out over the subcontinent and ultimately reached a great thickness – remnants today seen as the basalt of the Drakensberg Range.

Life during this fiery period must have been restricted to a few very hardy species, fossils of which have yet to be found. Of the marshes and lakes nothing remained. While life flourished elsewhere in the more hospitable parts of the earth, only the fossils sealed beneath the sterile basalt cap remained witness to the 50 million years when the Karoo was the world's Garden of Eden.

After the lava had ceased its flow, southern Africa experienced a period of uplifting; the supercontinent Gondwana of which it was part had begun to break apart. And while the massive 'islands' that were to become South America, Antarctica, Australia and India drifted across the seas, the central areas of the

120. (previous page) *From the top of the Nuweveld escarpment the vast sweep of the Park stretches across a panorama of plains and raised plateaux to the Grootswartberg range more than 100 kilometres distant.* **121.** *Even in drought years, spring sees the brief blooming of some of the Karoo's succulents. Here the delicate petals of* Hoodia bainii *contrast starkly with the ungainly, thorn-clad stems whose tops they coronet.*

122. *One of the most sluggish of South Africa's poisonous snakes, the puff-adder's danger lies in its reluctance to move from the rodent runways it haunts, and it is easily trodden upon by the incautious hiker.*

African subcontinent were elevated. The rivers no longer flowed inland but coursed to the seas, unleashing new and more powerful forces of erosion.

Deep valleys, destined to become wide plains, were cut into the Karoo rocks, exposing the alternating sandstone-shale structure so typical of the present landscape. Exposed, too, were the thick layers of basalt, the dolerite or 'ysterklip' which forms the resistant ledges along the tops and sides of so many Karoo koppies. And, brought back into the light of day, were the ancient, petrified skeletons of the long-gone Karoo animals, a thin sprinkling of bony clues to a mysterious and fascinating episode in the grand saga of evolutionary development.

Although, in the evolution of earth's animal and plant life, the story locked in the Karoo rocks represents but one very small chapter, the fossils have proved critical to our understanding of the evolutionary fortunes not only of the Karoo reptiles – but of the mammals and the creatures we know today.

The Beaufort West area, which includes the Karoo National Park, has become one of the Karoo's classic study and collecting areas. Indeed, it is likely that the earliest recorded Karoo fossil – a petrified tooth – was found at Beaufort West. The discovery of fossils here is largely due to work in the early part of this century by Rev. J.H. Whaits, who was without doubt one of South Africa's most talented collectors. An Anglican parson at Beaufort West with an intense interest in the Karoo's fossil history, he not only examined the flat plains round the town for fossils but also explored the Nuweveld

escarpment, an area partly incorporated in the present-day Park. These localities were also visited by South Africa's best-known and most versatile palaeontologist, Dr Robert Broom, who had come to South Africa in 1897 specifically to study the intriguing Karoo reptiles and their place in the evolutionary ancestry of mammals from reptiles.

And this study continues. In the Karoo National Park there is a clearly visible link between the geological horizons of Beaufort West, progressing through time layer by layer to those at the top of the Nuweveld escarpment. The deep ravines in particular reveal the alternating strata that accumulated when the Karoo was a marshland, while the thick caps of ysterklip are reminders of the fiery culmination of the Karoo episode.

There is of course the temptation to play amateur fossil-finder – indeed many of the early discoveries were made by interested amateurs. However, a word of caution is necessary. Fossils of any sort are of scientific importance and, as such, are protected in South Africa under the National Monuments Act, and a permit from the National Monuments Council is needed before specimens can be excavated. The regulations governing the collection of fossils are backed by the provision for heavy fines in cases of illegal excavation or damage. Fossils are surprisingly fragile and appear as white or yellowish bone fragments, sometimes in isolation, sometimes clumped together. Should you notice fossil remains within the Park, do not pick them up, try to dig them out or disturb them where they lie. Notify the Park authorities, for the find

has no value to science unless its exact locality is known.

But there is more to the Karoo National Park than a record of a distant geological past. The Springbok Hiking Trail traverses the plains and leads visitors to the highest point (1 900 metres above sea level) of the escarpment. In the course of three days and two nights of this trail they have a rare opportunity to experience the splendour of the Karoo; its vegetation (being studied and actively conserved within the Park); and a number of the larger mammals once associated with the area. Slowly being reintroduced here are animals such as the mountain zebra (related to the quagga which was hunted to extinction before the turn of the century), red hartebeest, gemsbok, springbok and that disconcertingly handsome creature the black wildebeest with its forward-twisting horns.

It is unlikely that the populations of these animals will ever reach – or be allowed to reach – their past numbers, but since the Karoo's many plant communities evolved under the pressure of grazing and browsing animals, the effects of these animals may well prove important in maintaining a varied array of plant-life here.

The Park is still new. Its special charms are not yet widely known, but as more and more people visit it and discover its stark and secret qualities, the poetry of its open vistas, the wonderment of its plant-life, the mysteries of its lizards and elephant shrews, baboons and porcupines, so the need to conserve and protect an even greater area of the Great Karoo will be recognised by the public whose support is so vital to the proclamation of future and larger national parks.

123. *The shales of the Park are of many colours and the toad grasshoppers which frequent a particular patch mimic its colours as protective camouflage.* **124.** *As the early morning sun takes the edge from the Karoo's chill night, small mammals such as these yellow mongoose emerge from their burrows.* **125, 126, 127.** *The ostrich, the world's largest bird, has long been a denizen of the Karoo, though its wilderness numbers diminished during the feather boom of the late Victorian era when farmers on horseback hunted the flightless birds to capture them live as breeding stock. The steenbok, too, (126) was once widespread in southern Africa, but as man has encroached on its habitat, numbers have fallen. Even the vast herds of springbok have gone. Here (127) a small group picks its dainty way across a stretch of scree, a sad relic of the past when, in the late 1890s, herds tens of thousands strong moved northward through these plains. To residents of Beaufort West the drumming of their passing hooves sounded as if a long drought had broken.*

123

124

125

126

128

128. *A kudu cow enjoys a succulent mouthful of late summer herbage in the safety of the Park's protection. These handsome antelope were at home in the Karoo long before the coming of the white man usurped and fenced their feeding grounds. Even today they disdain his boundaries and can clear a two metre-high fence in a graceful and seemingly effortless bound.*
129. *Blackfooted wild-cats, small bundles of spitting fury. Like the less pampered of their domesticated relatives, these nocturnal hunters prey on birds and rodents.*

129

THE ADDO ELEPHANT NATIONAL PARK

NATURE'S GREAT MASTERPIECE, AN ELEPHANT
THE ONLY HARMLESS GREAT THING.
JOHN DONNE. THE PROGRESS OF THE SOUL

'On the way down I visited the Addo, and soon realised that if ever there was a hunter's hell here it was – a hundred square miles or so of all you would think bad in central Africa, lifted up as by some Titan and plonked down in the Cape Province.' Thus wrote Major P.J. Pretorius whose name is part of Addo lore, in his 1947 book, *Jungle Man*.

A hunter's hell it may have been, but the accident of Nature that made the Addo Bush the impenetrable thorn thicket that it was – and still is – also made it the refuge of almost all the remaining elephant and buffalo in the Cape Province, but for one other small group of elephants which survives in the Knysna Forest in the Southern Cape, and it is on the verge of extinction.

As its name suggests, the Addo Elephant National Park was proclaimed first and foremost to protect the elephants of the Addo Bush. At the time of its proclamation, the people of South Africa were becoming aware of the losses being inflicted on their wildlife heritage as a whole, and of the horrific saga of the Addo elephant in particular. This concern had been translated into concrete form by the National Parks Act of 1926 which had provided the means; public awareness did the rest.

So the Park was established to preserve a single spectacular component of an ecosystem and not for the sake of the ecosystem itself. Though this runs contrary to modern conservation ideas, by protecting the Addo elephants the preservation of the whole complex was ensured.

The Addo elephants are spectacular and merit consideration in their own right. The tale of their decline in the 1920s is ugly, but their slower and eminently satisfactory recovery since 1931 is a refreshing success story and one with important ecological implications.

The vegetation that evoked in Major Pretorius 'all that was bad in central Africa' marks the zone where three distinct vegetation types meet – those of the Southern Cape, the semi-arid Karoo and the subtropical East African coast. It is a fairly low, evergreen and dense thicket of shrubs and small trees. Visibility in the Addo Bush is no more than a few metres, for the thicket ranges from 15 000 to 20 000 woody stems per hectare, creating a bewildering maze three to four metres in height. Add to this a formidable array of thorns and spines, and Major Pretorius's description takes on greater meaning. Other than a few streams that rise in the nearby Zuurberg and the Sundays River itself, the area had no permanent water. It also experiences gruelling summer heat, long periods of arduous drought, and rainfall that is at

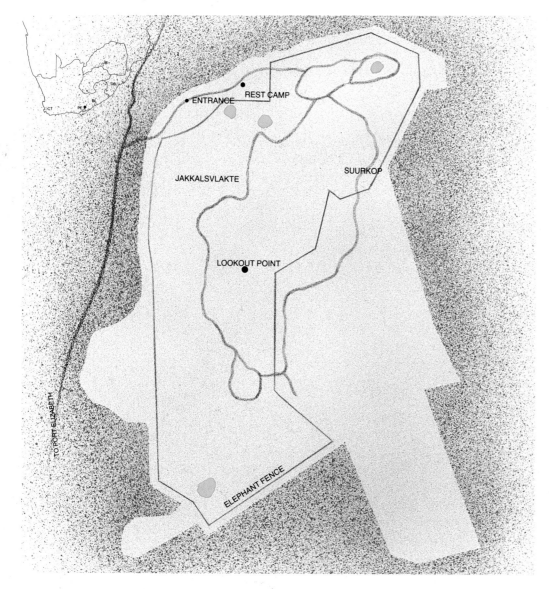

130. *A midsummer wallow.*

best irregular. It is not surprising that the Addo Bush was left to the elephants.

At the turn of this century the dark grey-green Addo Bush or Valley Bushveld – for this is what botanists call this vegetation type – mantled the low hills for some 50 000 hectares, reaching from the northern bank of the Sundays River and the foothills of the Zuurberg down to Kinkelbosch on the shores of Algoa Bay. In places the blanket of bush gave way to small clearings and seasonal pans, and everywhere elephant paths radiated from the pans and wallows into the secret heartland of the Addo Bush where these great beasts could rest, safe from ivory hunters. The elephant was not alone in this haven; buffalo, bushbuck, bushpig, duiker and grysbok also lived there and though hunters exterminated all the large animals in the open country surrounding Addo, those enclosed by the thickets were safe.

Their environment became an island cut off from the rest of Africa. No animals immigrated; isolation was total, to the extent that the great rinderpest epidemics in 1897 that laid low the game as well as the herders' cattle did not reach the Addo buffalo, huffing and blowing in the dust and thorns.

But time was running out for the 'old' Addo. At the centuries-old crossing of the Sundays River, known as the Addo Drift by whites (who had corrupted it from the Khoikhoi 'Gadouw'), a small settlement had grown up on the hunters' road from Port Elizabeth to Grahamstown. By 1875 a railway line from the coast to the interior snaked up the Coerney Valley, penetrating the Addo Bush. Less than 30 years later, just after the Anglo-Boer

War, the author Percy FitzPatrick was encouraging people to settle along the Sundays River which had become the site of an immense irrigation scheme planned to compensate for the poor rainfall.

Man was closing in on Addo. And for the elephants, water was the issue: to reach their traditional watering-places they had to leave the protective thickets, particularly during the dry periods, and in doing so they were increasingly harassed. Many were shot, many more were wounded when they broke fences on the way to quench their thirst at the newly-built cattle troughs and paused to sample the crops growing along the way. The consequence was inevitable; the farmers appealed to the authorities to destroy what they regarded as a menace to their livelihood.

The next episode begins with the smoke of guns and the stench of blood. In 1900, 150 elephants lived in the Addo area. Over the next decade, several renowned hunters tried their hand at exterminating them. All failed, beaten by the Addo Bush. Then in 1919 Major P. J. Pretorius was commissioned for a task somewhat less honourable than his exploits in the First World War. Of course this value judgement is based on hindsight; at the time, his commission to eradicate the elephants was accepted without question or qualm, and he set about doing so with the same skill and foresight that had distinguished his military career. Furthermore he brought to bear his experience as an elephant hunter and his superb marksmanship. In 11 months he shot 120 elephants.

This slaughter did not go unnoticed and press reports provoked much public

sympathy for the few survivors. Also increasingly influential was the fledgling Wildlife Society which campaigned to stop the slaughter. Sixteen elephants remained when the hunt was finally called off.

After much haggling and foot-dragging on the part of the authorities, Addo was proclaimed in 1926 a 'Provincial Elephant Reserve', but in fact offered little more by way of protection than had the earlier status of 'Demarcated Forest' – a status given this area in 1869. The hunt was ended, but the situation that originally had spurred it still needed a solution; the elephants continued to leave the reserve to find water. A borehole sunk for them was conveniently used by the provincial warden for his cattle. The reserve was unfenced and unpatrolled. The long-suffering farmers carried on where Pretorius had left off.

During this critical period the elephants found safety on the land of two remarkable men, Jack Harvey and his brother Natt, who allowed them to range unmolested on their property. Indeed it is possible that without their far-sightedness and tolerance, the Addo elephant would have disappeared.

The matter of the Addo elephants was, however, entering a decisive phase. After much public agitation and negotiation on their behalf, together with the intervention of the Anglo-Boer War hero Col Denys Reitz, the Addo Elephant National Park was proclaimed on 3 July 1931.

Only a small part (some 7 000 hectares) of the Addo Bush was set aside for the elephants, which were nearly 30 kilometres away on the Harvey land. They

131. *A rufous shag of hair on his face distinguishes the adult bull eland from the youngsters in the herd. A retinue of cattle egrets waits to gobble insects flushed from the undergrowth as the herd moves on.*   132. *Seldom seen because they prefer the densest thickets, the black rhino are nevertheless increasing in the Park. Using its prehensile pointed upper lip, this great pachyderm plucks a mouthful of thorny num-num.*

would have to be moved to their new sanctuary – a daunting task. The number of elephants was small – now only 11 were left – but they were unsettled and dangerous. There were no roads worthy of the name and there were no radios for passing messages to a line of beaters; gunshots and smoke would have to do. Moreover, the ill-tempered beasts would have to be driven through thickets so dense that the animals could not be seen until the last minute.

The first warden of Addo was one Harold Trollope, a man of considerable courage and ingenuity. He was not new to the wilds; he had been a ranger in the Sabie Game Reserve before this appointment. He realised that it was not enough to drive or coerce the elephants into their new reserve; he would have to ensure it was attractive enough to keep them there. Aided by his newly-recruited team of rangers, he set about providing permanent water supplies, and taking a firm hand with poachers. He also arranged a truce with the Park's neighbours. This moratorium was vital, for he had to make certain that no-one disturbed the animals during the critical phase when he started moving them into their new home.

Now began his epic elephant drive. To assist the droving team he recruited a local farmer, Piet Fourie. Using gunfire and tar smoke, Trollope moved the elephants. Where they crossed old trails leading away from Addo, fires were lit behind them to prevent them turning back. With initiative, cunning and a dash of sheer good luck, Trollope managed to drive the elephants into the Park with the loss of one bull he was forced to shoot at

close range when it turned on one of his men.

Then began a pattern which was to occupy the energies of wardens of the Addo Elephant National Park for the next 20 years – the ceaseless battle to keep the elephants from leaving the Park. This included repairing neighbours' fences, returning stock that had strayed through the breaks, and assessing the amount of compensation to be paid for cattle and goats killed or injured, for pumpkins eaten and maize fields trampled during the elephants' nocturnal wanderings.

Every morning they returned to the sanctuary of the Park. From time to time a wounded elephant would come home to die in the depths of the bush. It seemed unlikely that the elephant population would ever approach its earlier size. The eight births from 1943 to 1953 were discounted by eight deaths. Furthermore, relations between the Park and its neighbours were worsening until, finally, a deputation approached the government with an appeal for a satisfactory solution.

So many of the early strides in wildlife conservation in southern Africa depended on individual personalities; Addo's elephants were no exception. Graham Armstrong, soft-spoken and ingenious, believed he had the answer and set about building the world's first effective elephant-proof fence. The survival of the entire system encompassed by the Park was in the balance, for in the final

analysis any conflict between wildlife and man's economic well-being tips inexorably on the side of money and man.

After trying out various kinds of electric fencing, Armstrong became convinced that the answer lay in creating a barrier that even an elephant could not uproot. His prototype, built of scrap railway line and cables which were plaited by hand from strands of 8-gauge wire, was put to the test. On one side stood the elephants, on the other he placed oranges. He reported in triumph that his fence had even rebuffed a full charge by an elephant cow – it was truly elephant-proof.

At last it seemed that there was an alternative to exterminating the elephants to solve the man/elephant conflict. With financial assistance from the Wildlife Society and from the Port Elizabeth Publicity Association, and with the generous donation of old elevator cables from the firm Waygood Otis, an area of 2 270 hectares was fenced. The task was completed in 1954 – with 17 Addo elephants safely inside. In recognition of his efforts, the National Parks Board officially named the barrier the Armstrong Fence.

The elephants may have been the rationale behind the creation of the National Park, but they were not there in isolation. Nor could any serious planning be made for their future without a thorough account of the climate and vegetation. Showers fall here year-round, with a major peak between March and May and another not so pronounced one in spring (August). However, drought is no stranger to these parts, nor are periodic heavy rains and disastrous floods. The long-term rainfall averages 480 milli-metres a year. The oppressive heat of summer is somewhat tempered by cool afternoon sea breezes. Frost is rare, but early morning mists often hang softly over the valleys, the grey-green of the Addo Bush blurring upwards into the pale moist air above.

A closer examination of the Valley Bushveld – known locally as 'spekboomveld' because of the predominance of the spekboom – reveals one of South Africa's most productive habitats in terms of biomass. Biomass is a concept that allows comparisons between different environments, and it is simply the mass of animal and plant life a particular unit area such as a hectare or square kilometre can support. Thus in 1981 it was estimated that every square kilometre of Valley Bushveld could support a live animal biomass of just over 4 000 kilograms. In certain areas within Addo, estimates were as high as 6 726 kilograms, an astonishing figure exceeded in Africa only by some particularly rich habitats in East Africa. A fully grown elephant bull can weigh as much as 5 000 kilograms; hence a square kilometre of Valley Bushveld theoretically produces sufficient plant matter to sustain him. Of course, at this level of use the vegetation is suffering and is unlikely to maintain this biomass in the long term.

Sadly, outside the Park, the Valley Bushveld is under attack. Large tracts are being cleared for wheat and for pastures sown with alien grasses. The Addo remnant is thus becoming increasingly valuable. Today management of the area has moved from a total preoccupation with the elephants to a broader-based concern for the vegetation type – which is not only unique to this country, but of which Addo represents one of the last viable examples. Keeping this intact is a primary management objective, every bit as important as – if somewhat less exciting than – elephants.

All this points to a change in perspective, for if in the past the well-being of the elephants took precedence over all other considerations, today the elephants cannot be permitted to adversely affect the vegetation. There are elephants elsewhere in South Africa – indeed elephants are not yet endangered as a species. Should the Addo population be threatened it would have repercussions

133. *Africa has no rabbits but several species of hare occur on the subcontinent. Most come out at night and this handsome creature was spotted just as the sun set.*

in terms of genetics, as will be explained later in this chapter. For the vegetation the matter is entirely more precarious, and so in seeking equilibrium for the vegetation, elephants must take second place to the spekboom, to the boerboon and to a host of other plants.

Among the thicket of trees and shrubs, flowers enliven the prevailing grey-green: Cape plumbago, Cape honeysuckle and the climbing ivy-leaf pelargonium are known in gardens both here and overseas, but they are equally at home in the Addo Bush. When the boerboon comes into bloom it wears its brilliant red flowers like jewels – a few at a time. After good rains, the glades and occasional open areas in the Park dance with colour: daisies toss and nod, blood flowers thrust up bold-headed, chincherinchees provide a calm pale counterpoint to the veld vibrant with freesias, lilies and gladioli. Many of these plants, whose bulbs or seeds lie dormant beneath the soil during long years of drought or insufficient rainfall, suddenly appear and flower, laying up seed and nutrients to last until conditions favour them once more.

The birds of the spekboomveld are more often heard than seen. By day the sonorous call of the sombre bulbul, the sad liquid notes of the emerald-spotted wood dove, the exuberant call of the bokmakierie and the duetting of the boubou shrikes play through the leaves and stems. Of the 158 bird species on the

Addo checklist only 69 are permanent residents – the remainder are migrants or occasional visitors. Where the Karoo type of vegetation covers areas of the Park – areas such as Woodlands and Korhaanvlakte – birds as varied as ostrich and red-necked francolin, crowned guinea fowl and crowned plovers may be seen.

When darkness falls the haunting call of the Cape dikkop, the dull hooting duet of the spotted eagle owl and the plaintive litany of the fiery-necked nightjar provide the background to the black-backed jackal's howl.

Little is known of the insect life of Addo, but for oddities – such as the 'blue elephant stomach bot fly' – whose lives are intimately linked with the elephant. The larval stage of this particular insect must live in the stomach of an elephant, and despite the fluctuating fortunes of the Addo elephants in the past the insect has managed to survive. An even more noteworthy insect resident is the flightless dung beetle. The only species of a single genus, it is found in greatest numbers in the Addo Park, although it occurs sparsely in other areas of the Cape (and formerly was found as far afield as the Transvaal), where it has adapted to a somewhat more precarious existence living off cow dung. Because it has a predilection for the coarse dung of elephant and buffalo, it survives best where these animals occur in high densities. And its future is best assured in Addo which offers a consistent supply of food over a fairly large but confined area.

Today it is protected in the Park and is especially conspicuous after rain has moistened the earth. Then these beetles are encountered on the roads, rolling their balls of dung to some suitable spot to be buried. Their wings are vestigial and can no longer support the beetles in flight, so they must walk from one dung pat to another. However, because of their feeding habits, two of the beetle's six legs have become marvellously modified as rakes and scoops. Five delicate tarsi on the front legs – which is usual – would undoubtedly be a liability. Instead, the dung beetle's front pair of legs have evolved into stout, toothed paddles perfect for manipulating, patting and shaping dung into a ball.

Dung beetles of course, feed on dung. It is believed that the ball-rolling habit evolved as a means of removing a portion of the dung from the place where it is dropped and where a number of other dung-eating insects converge to take a share. By shaping the dung into a ball and then rolling it away from the competition, the dung beetle avoids having to tussle and squabble for food.

The dung ball is, however, more than a larder for the adult, it is also the perfectly cocooned and food-rich birthplace for the young. The female dung beetle lays her egg in a specially prepared 'brood ball' which she buries. Once the larva hatches from the egg it is not only surrounded by food, but the outside of the ball hardens, creating a protective case in which the entire developmental cycle of the beetle takes place until it breaks out ready to find its own dung pat and roll its own.

We know something of the flightless dung beetle. How many more insects, even other forms of life are dependent upon the entire Valley Bushveld system for their survival? We do not know – but so long as the spekboomveld, the elephant and the buffalo, the korhaan and the long list of named and as yet nameless species are protected, Addo offers a rich area for much needed research.

Before the system can be fully understood we will also have to understand the interaction and relationships between a wide spectrum of small vertebrates and their spekboom home. For instance, Tasman's girdled lizard is endemic to the region and the dwarf chameleon is also one of Addo's rare inhabitants. Though there are three species of tortoise and terrapin in the Park, only the large leopard tortoise is common, especially after rain when it becomes active, feeding on the new flush of herbs and grasses. But during dry periods even this species is not seen. Amphibians are, of course, limited by the lack of aquatic habitats but those species

which have adapted to long droughts, or which are independent of free-standing water such as the bushveld rain frog are well represented here.

It is likely that the small invertebrates, as well as the vegetation, birds and other smaller creatures were little disturbed by man when the Addo Elephant National Park was proclaimed in 1931. But this could not be said for the large mammals: they had been drastically affected by man and his works. The elephant remained – barely; the buffalo had fared somewhat better although it was nowhere as numerous as it had once been, living widely distributed over the wooded parts of the Cape Province. The Addo buffalo were the only survivors of these earlier great herds. That they had survived at all was a tribute to their adaptability. They had changed from being largely grazers to becoming browsers able to pluck the nutritious buds, leaves and twigs from plants such as the spekboom. Furthermore they had taken to skulking in the depths of the bush by day and only venturing into the open at night and then in small family groups.

Because this population has been isolated for so long it carries no diseases dangerous to domestic stock. This has had an important corollary: it is now possible to translocate buffalo from Addo to other reserves and parks throughout South Africa and even to neighbouring states without fear of their being a source of disease.

Within the dense bush the common duiker, grysbok, bushbuck and bushpig still lived but the kudu had disappeared. There were no carnivores larger than the

black-backed jackals – which are still common – and even today the caracal is rarely encountered.

The larger mammals were another matter and one which became one of the early management objectives – the restoration as far as possible of the larger mammal community. Eland, red hartebeest, kudu and black rhino had all lived here previously and so were reintroduced with ease. Unfortunately other species such as hippo, springbok and grey rhebuck, none of which are totally suited to the Addo environment, were also introduced. These species could never establish themselves and that they died out in due course proved a blessing in disguise.

Most of the introduced animals were brought from the nearest sources, but this was not so in the case of the black rhino. These cantankerous beasts were imported from Kenya in 1961 and 1962. As browsers they have found the habitat offered by the spekboomveld ideal and they have settled down and are breeding well in their new home. Nevertheless, bringing them from Kenya must be seen as a mistake, for the Kenyan black rhinoceros belongs to one of the northernmost subspecies of black rhino in Africa, while Addo is the former home of the southernmost subspecies, now extinct. The nearest southern cousins are found in Natal and the Transvaal. Today the mixing of subspecies is regarded internationally as being undesirable in terms of genetic variability. At Addo the mistake has been recognised – if somewhat belatedly – and later additions from Zululand have been removed from the Park.

Three species that are nocturnal, and thus not often seen, are the porcupine, bushpig and aardvark. But their existence is readily evident. The aardvark opens up termite mounds with its steel-hard claws to feed on the inhabitants. It leaves behind excavations that are later occupied by a host of creatures including the South African shelduck which regularly makes its nest in aardvark holes. As for the porcupine, its visiting card is the number

134. Seemingly raffish and nonchalant, a hefty mountain tortoise trundles down an Addo road. **135.** (overleaf) Major P.J. Pretorius called this landscape 'a hunter's hell' when in 1919 he was commissioned to shoot out the Addo elephants. Although he has the dubious distinction of killing six in 30 seconds, the Addo Bush won in the end. Dense, all but impenetrable in places, it provided the elephants with their last refuge. Here a family group lumbers through grey-green spekboom.

of striking black and white quills dropped along the road.

A surprising discovery has been that both the aardvark and the porcupine appear to conserve their food resources. The aardvark never demolishes a termite mound to satisfy its appetite, but excavates only a portion, culls the inhabitants and then leaves the survivors to repair the mound and rebuild their populations. The porcupine, it has recently been suggested, in nine cases out of 10 will leave behind a corm or tuber or two when it digs up a clump of bulbous plants for a meal. In doing so it loosens the soil and creates a shallow depression in which water can collect, improving the conditions for the remaining corm or tuber.

Although largely nocturnal, the bushpig is the most highly visible of the three. It plays an interesting role in the Addo environment, for it roots up relatively large areas, turning and aerating the soil as it goes, and in the course of its rootling creates ideal seed bed conditions for the successful germination and establishment of a number of plant species.

The Addo elephants' early reputation was not endearing for they had lived for many years in direct conflict with man: they were dangerous and mistrustful. Of those that remained within Armstrong's enclosure in 1954 at least two were regarded as man-killers, and for years wardens and rangers gave them wide berth – particularly the legendary bull 'Hapoor', named for a large notch in his left ear. The cows were equally

136. *Flaring its wings to protect itself from the writhing snake, a secretary bird pounds a prospective meal to death with its powerful feet.*

unpredictable and usually charged en masse if anyone entered the fenced area known as the 'Elephant Camp'. One neutral area remained, near the rest camp where the elephants were fed oranges during the winter months.

When biologist Anthony Hall-Martin began his research project on the Addo elephants in May 1976, he found this aggressive reaction, if inconvenient, a challenge. To work effectively he had to approach the elephants in the bush. It led to mutual acceptance, initiated by Hall-Martin who faced down a number of charges and in this way slowly won trust to the extent that his vehicle became accepted and he was able to drive up to the elephants without unduly alarming them. He describes the process: once the elephants had totally come to terms with his vehicle, other vehicles could enter the fenced area too, eventually making it possible to open the area to tourist traffic without fear of ugly confrontations between elephant and guest. Today the elephants take little notice of vehicles.

Feeding the Addo elephants is incompatible with the national park concept, but the original motive was sound enough. When Trollope took control of the Park the elephants were free to leave it in search of delicacies to which they had become accustomed during their forays into nearby orange orchards and pumpkin fields. Trollope was faced with a dilemma. He had, in a sense, to seduce the elephants from their former ways and his solution has a certain logic: what better inducement than the oranges for which they had already developed a taste? And so the practice of feeding oranges to the elephants began. Later it was found that visitors could safely watch this from a short distance and the feeding became a popular institution.

By 1977 the feeding issue had assumed new overtones. The number of elephants had increased to such an extent that competition between individuals and between family groups for the fruit was so intense at the feeding site that the animals had begun fighting to the point of injury. Artificial feeding was abandoned – to the relief of the purists. Nevertheless, it was an essential compromise at the time and was undoubtedly in the interests of conservation.

The Addo elephants were originally regarded as a separate subspecies but modern taxonomists place them with all other bush and savannah elephants of Africa, as a single species *Loxodonta africana*. However, a number of features differentiate the Addo population from others. Most striking is that the females

are usually tuskless, while the ivory of the bulls is notably small and light. This is a characteristic passed from generation to generation, and we can speculate on its origins in the genetic make-up of the population and perhaps also from the days when the biggest tuskers fell to the gun while those with more modest tusks were passed over – and went on to breed, presumably producing offspring as light in the ivories as the parents.

The ear and body shape of the Addo elephants is also somewhat different, but not sufficiently so to indicate that they are a separate species. Indeed, when one recalls that all the elephants at Addo today are descended from the 11 survivors of 1931 it is not surprising that they show the kind of mutual resemblance to be expected in a group isolated from other elephants and one which has experienced in-breeding. For many years the herd had only one mature bull, the infamous 'Hapoor', who almost certainly fathered all the calves born between 1954 and 1968 when he was shot after breaking through the Armstrong Fence.

His place as dominant male was taken by another bull 'Lankey' who fathered the calves born to the Addo elephants until more recent times when he had to contend increasingly with competition from other mature bulls. He, too, was eventually killed, by one of his fellows. The genetic stamp of these two bulls will be left on many future generations of Addo elephants.

Since 1954, the Addo population has increased at close to 7 per cent a year. Today there are 120 elephants and no indication that the birthrate will fall. And so the problem changes from saving the Addo elephants by building up their numbers, to one of coping with them now they are so numerous. The Park has only a limited area and already these great beasts are beginning to place pressure on parts of the ecosystem.

The situation well illustrates that the smaller the national park, the greater the need for management. Even though the Addo elephants are careful feeders, compared to their Kruger cousins who are notoriously profligate in their dining habits, the vegetation is beginning to suffer simply because it must sustain so many elephants.

Addo elephants do not push over trees, nor do they wantonly break off more than they can eat, but they are nonetheless choosy about what they eat. These preferences have led to at least six species of plant being totally eradicated or severely reduced. The elephants' sheer bulk has also had an effect; where they

137. *Addo's evergreen bush supports an incredible density of animals. These include troops of vervet monkeys which feed on its array of plants as well as on insects.*

rest and trample they open up the bush, allowing grass and other small plants to grow in areas previously shaded by thickets of trees. Of course this has been to the advantage of the buffalo and other creatures which prosper on such grazing. Inevitably, the elephants are bringing change to the unique plant community which is the Valley Bushveld.

This confinement of a large number of elephants to a relatively small area is also putting pressure on their own social fabric. One side-effect has been a dramatic increase in fighting between bulls, particularly as they mature and begin to cast covetous eyes at the females. Battling up the breeding scale has become so intense that the tussles may go on for days. If the Park were not fenced, weaker bulls could escape from the pursuit and attack of stronger males, but, unable to do so, they must take the consequences. More are injured and killed in such conflicts at Addo than in any other known elephant population.

What of the future? For the third time

this century it is necessary to find a solution to the Addo elephant problem. Will it be found in the crash of a heavy rifle or the whisper of a drug-filled dart, or will there be an alternative, as Trollope and Armstrong offered alternatives to the Pretorius solution?

There are good reasons to believe that 500 animals is the minimum population necessary to ensure adequate genetic variability. To support 500 Addo elephants, a minimum area of 40 000 hectares – with the same carrying capacity as the existing Park – would be required, or a larger area of lower nourishment potential. Thus the key to the Addo elephants' future lies in more land. It is also crucial to the future of the Valley Bushveld. To relieve pressure on the vegetation, in 1977 the Armstrong Fence was extended to enclose 3 953 hectares. By early 1982 a third phase of fence construction brought the elephant range to about 5 127 hectares. The fencing team is still working but even when the rest of the park is enclosed, including the

800 hectares acquired in 1981, this will still only be enough for 90-95 elephants, less than the present population. Yet the numbers continue their upward spiral.

Translocation is not the answer to such pressures, for while areas such as the Andries Vosloo Kudu Reserve north of Grahamstown could provide for a number of them, there are no areas large enough to carry a viable population and once again the 'magic' number 500 comes up: to maintain the Addo genetic mix and some guarantee of its survival, future planning will have to provide for one large elephant population in the area. This applies to the buffalo as well. At present they number between 100 and 300 animals, depending on conditions; for them 500 is also a genetically safe population figure. But if they are to increase to this, they too will need more land.

There is one other alternative: culling. The National Parks Board is reluctant to take this step, for it further jeopardises the genetic make-up it seeks to protect. Land is expensive. It has other uses financially more profitable than 500 elephants. Farmers have introduced the Angora goat to the spekboomveld; these, too, flourish on the rich browse, and at shearing time each year they turn a profit which no elephant can equal. With the help of a recent cash donation the Board purchased 800 hectares of land adjoining the Park, but it is not nearly enough to even catch up with the elephants' present needs.

Elephants are emotive animals and perhaps, in due course, the public will respond to the need for funds. If it does, it will ensure another immense benefit – the spekboomveld in the National Park will be protected. Once more, by rallying to the needs of the elephants we will also be serving the interests of the boerboon and the dung beetle and the myriad smaller creatures which need protection but lack the glamour to attract it.

138, 139. *Rolling a ball of buffalo dung, the wingless beetle,* Circellium bacchus, *typifies the many smaller forms of life protected within South Africa's national parks. Ball-rolling, it has been suggested, allows the dung beetle to remove a portion of its food supply, thus avoiding competition with several other animals such as these red-necked francolins (139) foraging for seeds. But dung is more than just food to this beetle; once safely buried, the ball provides a secure cocoon in which it lays its eggs and the young develop, feeding off the contents of the ball.* **140.** *Anyone venturing on foot through Addo's concealing bush risks coming upon the awesome Cape buffalo. Possibly because they were hunted in the past,*

138

139

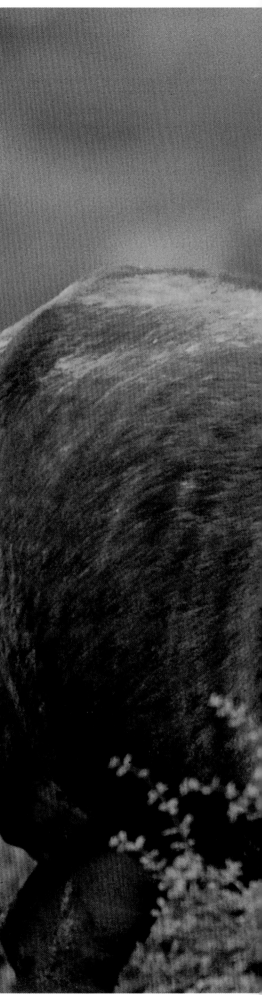

140

they lead a largely nocturnal existence, venturing from the sheltering thickets only very early or late in the day. The elephants' feeding habits, their pathways and trampled resting places create clearings which are changing the vegetation of Addo – allowing grass to spread and so, in turn, encouraging the Cape buffalo which prefers to graze rather than browse, to increase as well.

142

143

141. *Refreshed by a mudbath, two young bulls engage in a bout of sparring. Their small tusks are typical of the Addo bulls; the cows are completely tuskless. At one time it was believed that they, together with that other last southern remnant, the elephants of the Knysna Forest, were a separate subspecies. Today it is accepted that they are the same species as Africa's other elephants and that differences of ear shape, generally smaller size, and lighter ivories are more likely to be a combination of genetic variations and environmental adaptations. However, research into the question continues.*   **142.** *The elephant's great mass makes it difficult for it to lose excessive heat. Here one takes a cooling plunge into a wallow.*   **143.** *The elephant's enormous ears help it cope with the build-up of heat. By circulating much blood through their ears, wafting and wetting them, these massive animals cool off.*

144

145

144. *Family herds are led by females, but the bulls constantly compete within their hierarchy. Only the dominant males win access to the breeding females and over the past 30 years first the infamous Hapoor and later, after he had been killed, another bull, Lankey, sired most of the calves born in Addo. Here two males have skirmished, and the loser submits, allowing the winner to mount him in a mock mating posture. Because of the build-up of the elephant population from an original 11 to more than a hundred today, social pressure is evident and the Addo elephants fight more than any other known population in Africa – sometimes to the death.* **145.** *A young female challenged by the matriarch of a rival herd at a waterhole, scampers off to join her own.* **146.** *No more than a few months old, this baby may well live until its teeth wear out at about 60 years of age. After half a century of protection the Addo elephants continue to breed successfully, the average Addo cow producing a calf every three to four years. Today these elephants face a new crisis as they threaten to outgrow and destroy their Park.*

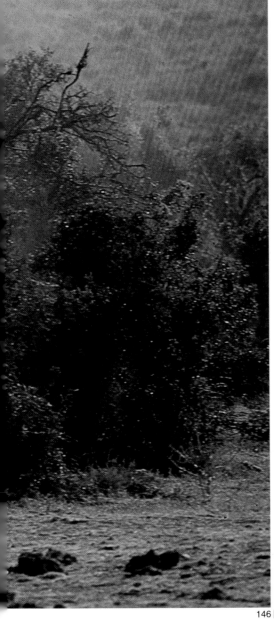

THE MOUNTAIN ZEBRA NATIONAL PARK

HARNESSED WITH LEVEL RAYS IN GOLDEN REINS,
THE ZEBRAS DRAW THE DAWN ACROSS THE PLAINS
ROY CAMPBELL. THE ZEBRAS

On the basis of its scientific name *Equus zebra*, the Cape mountain zebra takes pride of place in the stable of African striped horses. Yet of the three existing species (the fourth, the quagga, is now extinct) and two subspecies of zebra it is the only one listed in the Red Data Book of endangered species as being 'vulnerable'. As recently as 1930 it would have justifiably earned a place as 'endangered' – if not as 'rare'. Its gratifying change in status has been a direct result of a concerted effort by people deeply committed to saving this attractive animal from the same fate as its cousin, the quagga.

Initially its plight went unrecognised – in the eyes of the general public a zebra is a zebra and the differences of stripes and habitats make little impact. The fact that Burchell's zebra was so clearly thriving on the African savannahs satisfied most people and tended to obscure the fate of its relations living only in the mountains of the Cape Province south of the Orange River. Yet they are quite different: unlike Burchell's zebra the mountain zebra has no shadow striping between the main black stripes, has a distinct dewlap, a thinner mane, an orange muzzle, white stomach, a characteristic grid-iron pattern on the rump, and the dark stripes continue all the way down the legs. It is also smaller and specially adapted to rocky terrain.

The mountain zebra was not without allies, and early in the 1930s the National Parks Board recommended that a farm in the Cradock district be bought in an effort to protect this rare animal.

But in the minds of many conservation was still something of an expensive luxury – particularly when it involved paying for land. The Minister of Lands at the time stopped the purchase, under the false impression that the mountain zebra at the Cape and the somewhat larger (and browner) Hartman's mountain zebra – in what are now Angola and Namibia – were identical. Since Hartman's mountain zebra was fairly numerous, he saw no reason to pursue the plan.

Efforts to explain the error went unheeded until the voice of Chief Justice de Wet added authority to the increasing clamour. In a personal letter to the then Prime Minister, J.B.M. Hertzog, he urged that options be taken on certain farms in the Cradock area to save the Cape mountain zebra. The issue could no longer be pushed aside and in 1937 the

1 712-hectare farm Babylons Toren was purchased and proclaimed the Mountain Zebra National Park. Since then more farms have been added and today the Park encompasses 6 536 hectares.

From the start the main purpose of the Park was to preserve a viable, genetically pure population of the species. The prospect of success was not encouraging; nor is the present situation any guarantee, for it is generally accepted that 500 of a species is necessary in one area to ensure long-term genetic diversity and thus a strong breeding stock. The carrying capacity of the Park is only 200 mountain zebra, a fact which must affect future planning. If the Park is to attain its primary aim, more land is needed to allow the number of mountain zebra to breed to the 'safe' limit.

While today's scientists work towards the magic number of 500, the Park's early managers must have wondered how they were to save the mountain zebra at all. When proclaimed, the Park contained five stallions and a single mare which died a year after producing a filly. By 1945 only two of the stallions and the young mare remained. A year later the mare died without having produced offspring. With two stallions as the sole representatives of the species the Park was expressly intended to protect, the future seemed bleak.

However, the mountain zebra was not yet extinct. Several still existed on nearby farms and in 1950 five stallions and six mares were driven into the Park. By 1964 this original core had settled down and increased to 25. To it were added 30 from

147. *Showing their distinctively striped hindquarters, mountain zebra step nimbly up a stony slope to graze the sweet grasses on the Park's Rooiplaat plateau.*

148, 149. *Predator and prey in the rocky haunts of a Karoo koppie. This adult dassie (rock hyrax) – surrounded by youngsters – kept a careful watch for danger from black eagles in the sky and other predators on the ground. At the first sign of danger she was ready to shoo her young into their nearby burrow. Danger was close, for this caracal (149) had spotted the family group. Moments later it leapt into the open, scattering the terrified dassies. Bounding, turning, quick and keen the cat pounced, its full weight brought to bear on a hapless dassie, and the hunt was over.*

yet another farm, bringing the total to 55, and the Park was extended to its present size. Since then the numbers have built up steadily so that today animals in excess of the 200 judged sufficient for the Park are translocated. In the past seven years 83 mountain zebra have been moved elsewhere, to create three other breeding populations: in the Karoo Nature Reserve at Graaff-Reinet, in the Tsolwana Game Park in the Ciskei and in the Karoo National Park at Beaufort West. Here 30 animals have been introduced and appear to be thriving, not surprisingly, for the Mountain Zebra National Park and the Karoo National Park both incorporate mountain plateaux and slopes clad in the sweet grasses on which this zebra flourishes.

Like its sister park to the west, the Mountain Zebra National Park is distinguished by a degree of aridity, spectacular daily fluctuations in temperature, and the trying contrast between winter when snow shrouds the Bankberg and the valleys of the Park, and summer when temperatures soar to 42 °C and strong mountain winds scorch the plains. Yet of the two parks, it has a kinder aspect.

Lying in the great amphitheatre formed on the northern slopes of the Bankberg, it enjoys about 400 millimetres of rain a year. The Wilgerboom (Willow) River with its well-wooded banks runs the length of the Park, and the topography includes mountains, plateaux and low-lying valleys.

An arm of the Park, known as Rooiplaat, pushes north to incorporate an extensive plateau of sweet grasses on which most of the larger grazing mammals can be seen. Here springbok, blesbok, black wildebeest and red hartebeest were introduced; five years ago kudu moved into the Park of their own accord.

The mountain reedbuck has not only thrived on the rugged slopes, but today its numbers must be controlled by culling or translocation. It ranks second to the mountain zebra as a specially conserved species in the Park. Here, too, the massive eland can be observed picking its way over the stony terrain.

The mountain zebra is usually seen in small groups comprising a stallion and two or three mares with their foals. Young males and adult males without mates tend to group together in small loose herds. This species of zebra is not territorial but a stallion will protect his mares, particularly from bachelors keen to woo them away, and there are dramatic clashes of rampant males fighting to possess the breeding females.

When the Park was proclaimed certain antelope such as the grey duiker, steenbok, klipspringer and grey rhebuck were already there, though never in any great number. However, free of their natural predators, the larger herbivorous mammals have flourished – and in doing so pose a new problem. Long ago lion and hyaena were killed off by farmers anxious to protect their stock, and today the largest predator in the Park is the caracal, which can grow to 20 kilograms and is quite capable of bringing down a mountain reedbuck.

Thus the success of the Mountain Zebra National Park in many ways has become its threat. It was intended that once the mountain zebra was secure, the Park should support as wide a spectrum as possible of its former inhabitants and these were reintroduced. However, contained by fences, and deprived of the balancing influence of predators, the numbers have built up alarmingly. Rooiplaat, for instance, has been seriously over-grazed in the past.

It has been suggested that a sound basis for the utilization of habitats would be: 'more should not be eaten during a given period than can be produced during a period of rest.' In order to allow this to happen, the Park managers have where possible had to translocate animals considered in excess of the carrying capacity of the Park. Where this has not proved possible, excess numbers are culled, though notably not in the case of the mountain zebra.

Animals such as blesbok, black wildebeest and ostrich have also had to be removed. Just as the future of the mountain zebra needs the guarantee of more land, so the other animals that share the Park would benefit and plans are afoot to increase its size.

An exciting aspect of the Park is its bird-life, for this arid area with its rocky heights and numerous reptiles and small mammals, abundant insects and frogs, is frequented by some 200 species, of which three are listed in the Red Data Book. On the open plateaux such as Grootmat and Rooiplaat, the black harrier lives most of the year. Once believed to be very rare, it can be seen soaring high over its breeding territory exposing its distinctive white rump and splendidly banded wings.

Equally rare is the Cape eagle owl, of which remarkably little is known. The third notable species found in this Park is the booted eagle. Two populations of this bird occur: one breeds in Europe and western Asia and migrates to the African subcontinent in the northern winter; the

150

151

other breeds here in the austral summer and migrates to Angola during the winter months.

Along the paths of the three-day Mountain Zebra Hiking Trail, the visitor can discover the wonderful succulents and grasses typical of the area, scramble up ravines overgrown with wild olive, and pause beside mountain streams and pools so lushly grown with ferns they look like tropical hideaways. On foot or from a motor-car, the visitor experiences the marvellous contrasts and unusual beauty of this small but intriguing portion of the Great Karoo.

150. *Female mountain reedbuck on the Park's rock-strewn slopes. This antelope is second only to the mountain zebra as a specially conserved animal here, and its numbers have increased greatly in the sanctuary of the Park.* **151.** *Compared to herbivores, flesh-eaters have a harder time in filling their bellies and all of them – the lion no less than the humble Cape fox shown here – are opportunists, scavenging when they must.* **152.** *The view north from lofty Kranzkop across the rugged terrain of the Park to the table-top koppies and plains beyond. On a March day in 1976 a 6 000-tonne rock tumbled 200 metres down the mountainside leaving as its wake a massive groove. This scar is fresh and clearly visible in the Park but the surrounding countryside reveals evidence – now blurred by erosion and plants – of similar such landslides that have helped sculpt the Karoo's characteristic landscape.* **153.** *Forty centimetres of sluggish reptile soaking up the morning sun, a rock monitor draws on midwinter's early warmth to heat its body before it can begin its daily hunt for insects, birds' eggs and other small prey.* **154.** *Each bloom of this wild morning glory lasts little more than a day; tomorrow it will be replaced by a new one.*

152

153

154

155

156

157

158

155. *Snow regularly blankets the winter Park. Here a blade of prickly pear pokes from a recent fall. Introduced many years ago from the Americas, these cacti have now spread widely in this country but are being eliminated from the Park in keeping with the policy of removing all alien flora and fauna from South Africa's national parks.* **156.** *Bleak winter landscape.* **157.** *Ostrich foraging in snow on high-lying Rooiplaat plateau.*

158. *Mountain zebra seek grazing in the snow-clad landscape of Rooiplaat plateau. From the first five stallions and six mares driven into the Park in 1950 – and augmented by a further 30 some years later – the mountain zebra has flourished, and is now less likely to join its cousin the quagga in extinction.*

THE GOLDEN GATE HIGHLANDS NATIONAL PARK

THE BIG ANTELOPE RACE UP FROM THE PLAINS. . .
STRAINING WIDE THEIR NOSTRILS
AND THEY SWALLOW THE WIND.
EUGENE MARAIS. THE DANCE OF THE RAIN (TRANSLATED BY UYS KRIGE)

Enjoyment is important to our appreciation of all the national parks but in the Golden Gate Highlands National Park the recreational and aesthetic are foremost. In many ways, compared with parks such as Kruger and the Kalahari Gemsbok, it is the odd man out: where in most of the others the landscape is regarded as secondary to the ecosystem it contains, to the wildlife and vegetation, in this Park the scenery and setting are its *raison d'être*.

Here the visitor will not wipe dust-streaked sweat from his face and experience the adrenalin charge of a lion kill, nor will he sleep in a rest camp fenced for his safety. Where the other parks are active, the Golden Gate Highlands National Park is passive. Its unspoilt wilderness evokes the contemplative and engenders a sense of order and peace in the thousands of visitors who make the few hours' drive from the heavily urbanised Pretoria-Witwatersrand-Vereeniging complex.

This proximity to the industrial heart of South Africa was well in mind when in 1962 the Parks Board, at the invitation of the Administrator of the Orange Free State, set about investigating a Highveld national park. While inspecting several possible sites, almost by chance the Board

members visited the foothills of the Maluti Mountains on the border between South Africa and Lesotho and there found a region of Highveld spectacularly combined with mountainous scenery in which gigantic outcrops and cliffs of stratified sandstone, marvellously eroded by wind and water, dominate grass-swathed hills and sweetly chuckling streams and rills. Entranced by this singular beauty, the Board made its recommendations and in September 1963 the Park was proclaimed expressly to preserve the landscape.

Just over 6 200 hectares in extent, it contains a fairly large and representative

section of the Cave Sandstone formations that lend the Park its rugged appeal; at the same time it protects the source and upper catchment area of the Little Caledon River. Another attribute is the rich alpine-like sourgrass veld conserved here, as well as the glorious flowering plants – and the lammergeyers soaring high in the sky.

The air has the pristine crispness of the mountains, and as the Park is both at a relatively high altitude and well inland, the seasons are better defined than elsewhere on the subcontinent. In the Golden Gate Highlands National Park visitors can enjoy snow in winter followed by the tender green of spring, the

159. *Blazing richly against a cloudless sky, towering sandstone cliffs do justice to the 'Golden Gate' from which this Park takes its name. Over tens of thousands of years, the oxidization of minerals within the strata have stained the rockface with a rich patina of golds, coppers and reds, while wind and water have sculpted shapes into likenesses limited only by the viewer's imagination. Nature's carving to the left has been dubbed 'Gladstone's Nose' while to the right are the 'Mushroom Rocks'.*

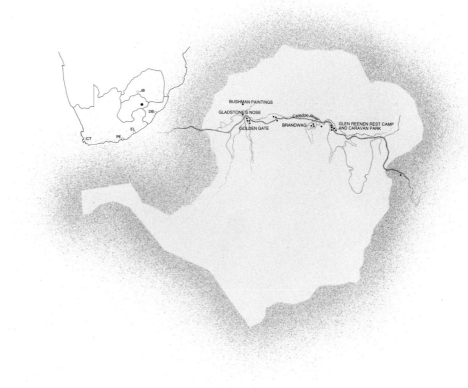

strong ochres and warmth of summer fading into the full autumnal palette as winter descends once more.

And when the seasons change the pigments, the rock formations provide the special qualities of the canvas. Geologists describe this as part of the Karoo Supergroup which embraces much of central South Africa. In the eroded overhangs and caves, the upthrust plateaux, and in the Maluti Mountains, the geological layers are revealed as three distinct strata: topmost is the basalt which forms these mountains as well as those of the Drakensberg and was laid down by immense lava flows, subsequently uplifted and eroded into the spine of mountains parallel to the eastern seaboard of South Africa. In the Park the basalt layer occurs above 2 300 metres and yields unusually fertile soil. Indeed, scientists believe that these soils may be the most fertile in the entire country. They sustain dense temperate grassland which completely protects the upper reaches and steep high slopes from erosion. Below this is a layer of Cave Sandstone, also known as Clarens Sandstone taking its name from the tiny village 20 kilometres from the Park which, in turn, derived from the town in Switzerland where Paul Kruger – the founding father of the National Parks in South Africa – spent his exile.

Cave Sandstone is relatively easily eroded, but the layer below it, the Red Bands of reddish-brown mudstone, is even softer and together they account for the formation of wondrous shapes such as Baboon Krantz, Mushroom Rock and the whimsical Gladstone's Nose, for the rocky sentinels, numerous caves and craggy

overhangs, for the strange shapes that nudge the imagination – does it look like a frog? a moon monster? a disgruntled aunt?

Soils from these two layers tend to be infertile and powdery and the vegetation they support is therefore poor. In the years when the Park's area comprised seven privately-owned farms, overgrazing by domestic herds on these lands led to increased erosion and farmers were distressed to find that the arable valleys and lower slopes were gradually being overlaid by these impoverished soils washed down increasingly by the rain.

The farmers were not the first people to settle here, for the innumerable stone implements, cave dwellings and ancient middens reveal long human occupation. And, long before man trod this earth, dinosaurs inhabited the area. This is confirmed by fossil finds that link with the same mammal-like reptiles and vegetation that makes the fossil beds of the Great Karoo of such immense interest.

Current theory places the Bushmen (San) as the first figures in the southern African landscape and, while the primitive paintings daubed on the walls of several caves in the Park have been dated to as recently as 200 years ago, there is little doubt that bands of these diminutive apricot-complexioned people sheltered in the caves and overhangs, and hunted the wild herds here for many thousands of years.

Later, some 1 500 years ago, the Bantu-speaking people arrived in the foothills of the Maluti Mountains. A taller race, darker skinned and armed with iron-tipped spears and arrows, the men drove their herds of small, wiry cattle across these same valleys in search of water and pastures; the women followed, iron hoes slung over shoulders.

They were part of a great slow migration that pushed southwards until it reached its southern limit in what is now Transkei – perhaps as recently as 200 years ago – and doubled back on itself. Suddenly there was no more land to settle, no more land on which to spread out the herds. Populations built up, water and grassland were in increasingly short supply, and there rose up powerful chiefs who, with their well-armed military, set about dispossessing their neighbours.

The foothills of the Maluti Mountains were not unaffected by these times of fear. When the first white Trekkers in search of an independent homeland reached the area, they found ownership of land in dispute and the local people unsettled and dispersed.

The arrival of the Trekkers at the Golden Gate is not strange for it is indeed a gateway from what is now the Cape and Orange Free State to the Transvaal and to the Indian Ocean coast further to the east. Piet Retief, leader of the Great Trek, is said

160. *Guarding the western entrance to the Park are the twin sandstone buttresses of the 'Golden Gate' through which the Little Caledon River flows below the Little Golden Gate Dam in the foreground. The buttresses' siting is fortuitous for this was a 'gate' not only in appearance, but one through which countless waggon trains of Trekkers passed.*

to have outspanned here. And in the course of a six-day hunt in the autumn of 1837 his men trundled back to the laager nine wagons straining and creaking with their burden of meat for provisions, and skins taken from widely varied and numerous game animals.

It must have been a hunter's paradise, for another slightly later report states that one Cornelius Roos and his sons shot more than 300 lion on and near his farm which lay between Golden Gate and the nearby town of Bethlehem.

When proclaimed the Park no longer carried any of the larger mammals – predator or prey. Apparently the only animals of note were some mountain reedbuck, grey rhebuck, baboon and black-backed jackal. On all seven farms subsequently included in the Park, the farmers had killed the predators as vermin. The antelope had fared little better. They had been shot for the pot and because they consumed grasses which could be fed to cattle and goats. Another factor that undoubtedly contributed to their demise and to the demise of wild animals throughout the world, was the impact of agriculture; as the land was cleared and planted, divided up and fenced, so the game lost its feeding grounds.

It was not long before this crossroads to the hinterland was tamed, cowed, transformed.

Soon after the Park was proclaimed scientists began a study of its natural resources. They listed and observed, unearthed and pondered on the trees, the flowering plants and grasses, on the soils, fossils and strata, hand-axes and cave-paintings. They also tried to establish what animals had lived there in the recent past, deducing this from a study of bone remains, from the Bushman paintings, and from accounts by the Trekkers. From these scattered sources, a picture emerged of a land abounding with seasonal game. In certain months of the year there would be Burchell's zebra, black wildebeest, mountain reedbuck, steenbok, blesbok, red hartebeest, grey rhebuck, klipspringer, oribi and baboon. Of the larger predators there were lion, leopard, hyaena, wild cats and Cape hunting dogs, as well as the lammergeyer which alone, of all this company, remains.

The Park has been restocked with all the herbivores that formerly inhabited the area, and several have thrived, even though it is likely that a number of the larger antelope would have been migratory rather than resident. Within the Park visitors can view these animals – including the largest herds of black

wildebeest in the country. The habitat has also proved ideal for the oribi, a somewhat uncommon small antelope often mistaken for the smaller steenbok.

Unfortunately none of the predators has been reintroduced for not only is the Park too small to accommodate fully functioning ecosystems, but this would conflict with the recreational aspect which is part of the rationale for the Park, a rationale which places it outside the generally accepted parameters for a national park. But its role as a means of increasing public awareness of the value of wilderness and its conservation offsets this deviation from the norm.

Visitors to the Park have a choice of accommodation, from luxury bungalows to caravan facilities and simpler but no less comfortable rondavels. There is also a nine-hole golf course, tennis courts, swimming pool and a stable of horses.

But it is the outdoors that holds the attention. You can stroll along paths from the two-day Ribbok Trail to shorter rambles with names such as Boskloof, Sentinel, Wodehouse Peak, Holkrans, Echo Ravine and the Lammergeyer Trail.

Close to the Glen Reenen rest camp, the Lammergeyer Trail passes near the place where a pair of lammergeyers used to nest. Although this raptor has an extensive range through Eastern Europe and across Asia to China, and is known in East Africa and in Ethiopia (where there are an estimated 16 000) only a handful survive in the mountains of South Africa. This has not always been so and there are accounts of lammergeyers nesting on Table Mountain in the Cape Peninsula at the southern tip of Africa. But over the last 300 years, since the beginning of European settlement in southern Africa, its fortunes have tumbled drastically. This decline can be explained largely by the bird's specialised diet of bones, bone marrow and, to a very much lesser extent, rock hyraxes or dassies.

The main ingredient, bones, has become more and more scarce and where

the lammergeyer, like the hyaena, could depend on a goodly supply of skeletal remains after the other predators and scavengers had taken their fill, today the veld provides a scant larder. Carcasses are no longer common and the pickings for the lammergeyer have grown increasingly lean and its population dwindled.

The lammergeyer or bearded eagle (from the distinctive bristly tufts about its beak and chin) has evoked much controversy, some ornithologists claiming it to be a vulture. But it is an eagle, albeit a vulture-like eagle, and furthermore the largest in South Africa. No one has explained the function of the 'tuft' but the renowned ornithologist Leslie Brown suggests that it may serve to warn a bird poking its long beak into a bone in search of marrow that it has pried deeply enough and is at risk of getting stuck.

Marrow and dassies contribute to the bird's fare, but it is more famous for its prodigious bone-eating. An observer once watched one swallow the thigh bone of an adult zebra. Such gastronomic excesses are digested by the bird's extraordinarily powerful stomach juices that break them down completely. Unlike the owl, the lammergeyer does not regurgitate indigestible fragments but joins the hyaena as the only creatures equipped to handle bones, earning for the bird the sobriquet 'avian hyaena'.

The Greeks were the first to record the bearded eagle's unusual habit of carrying bones to a considerable height and then, with amazing accuracy, dropping them to shatter on a rock far below. They accordingly named the bird 'ossifractus' – the bone breaker. The birds have favourite dropping sites which are veritably littered with bone fragments.

Pliny the ancient poet tells a salutary tale of the lammergeyer already known for its habit of shattering the carapaces of live tortoises by dropping them from a height. His friend Aeschylus heeded an oracle that he would be killed by the 'collapse of a house'. The playwright took this

literally and spent the day out of doors – to die when a passing lammergeyer dropped a tortoise on his bald pate.

The pair which until a few years ago used to breed opposite Glen Reenen rest camp built their large nest of branches lined with sheep's wool and fluff. They also had another nesting site in the vicinity and, while they did not use it, they maintained their rights to it and kept other birds away. In the breeding season the female (the pair remains together for life) lays one or two eggs and after about eight weeks the first chick appears, followed a few days later by the second. If both chicks hatch, the second hatchling soon disappears, apparently the victim of fratricide – a practice common among raptors. Why this happens is uncertain. It has been suggested that the second egg is merely insurance should the first fail to hatch and that one fledgling is all that the adults are able to raise. This is difficult to prove, for irrespective of the availability of food, only one chick remains. It seems that the older, and thus more vigorous chick takes the major share of food brought by the parents to their offspring and that the younger becomes increasingly persecuted and weakened to the point where it may even starve to death.

As the remaining chick matures its fluffy feathers are replaced by its full flying plumage and in time, the voracious youngster is ready to become airborne, leave the nest and its parents' territory. A male will have to seek a territory of its own if it is ever to attract a mate and breed; a female will have to choose a mate.

The landscape that the lammergeyers make their territory, including the Golden Gate Highlands National Park, is not only transformed by the seasons, but is adorned with the brilliant colours and forms of its myriad flowering plants.

Here grow graceful kniphofias, wild gladioli, agapanthus with their cool blue flowering panicles, watsonia, the

assertive spikes of red-hot pokers, and an
array of buttercups, everlastings,
dandelions and aloes. And all about are
the grasses, lush and burnished by
summer, green and tender in spring, dry
and gold as winter approaches.

The Golden Gate Highlands National
Park is of undoubted beauty, but it is a
beauty that can achieve more than just
pleasure: it can teach. This is the only
park in the country to have a permanent
school which children attend for short
periods to learn of the principles of
conservation and observe at first hand the
wonders of the natural world. And if
every child who attends this special
school, and every person who walks the
trails, or stands entranced as a butterfly
struggles damp from its pupa, or as the
lammergeyer swoops low and fast over a
sandstone ridge, as a wildebeest kicks up
its heels or as the sun plays across the
cliffs and gullies, feels suddenly his
origins and acknowledges his or her
responsibility to maintain some link with
them, then this Park will have made its
contribution to conservation.

161, 162. *Until less than two centuries ago the
hunter-gatherer Bushmen (San) made their
homes in the rock shelters which pock the
cliffs of the Park. Today their only traces are
rock paintings. Was it for some form of
totemistic magic that the long-gone artist
evoked these eland (161) without horns? Later
settlers dispossessed the Bushmen and
decimated the eland, though these (162) have
been reintroduced to the Park.* **163.** *The
towering majesty of Cathedral Cave with its
30-metre cascade, testifies to the awesome
force of water. Here, over countless aeons, the
stream has carved its persistent passage deep
into the rocky folds to create one of the Park's
most magnificent spectacles.*

163

165

166

164. *A cave mouth on the Holkranz Trail frames the vista of plains and mountains specially conserved in this Park.*   **165.** *Stamping agitated hooves in almost heraldic splendour, the black wildebeest patrols his territory on one of the Park's grassy plateaux. With the advent of early summer he is about to shed the winter coat which has protected him from the often-freezing winds and falls of snow. Forward-pointing horns and a conspicuous white tail distinguish this species from its cousin the blue wildebeest.*   **166.** *Sated on the remains of a blesbok carcass, Cape vultures lift off in ungainly flight to resume their holding-pattern patrols above the landscape. Rising on thermals to great heights, they keep constant watch for any likely carrion as well as on each other so that when one scavenger drops from the sky it is soon followed by others. In minutes a carcass becomes an avian shambles with pecking and clawing birds everywhere.*

167

167. Fire sometimes brings its benison to the grasslands of the Park. Here, against a sculpted sandstone backdrop, a herd of Burchell's zebra graze the new shoots of summer which have emerged from the ashes of a fire sparked by lightning.   **168.** Since the inception of the Park several species which once made their home here have been reintroduced. Among them are blesbok, seen here at a canter as they breast a rise.

168

169

169. Throughout the Park the many man-made dams attract flocks of waterfowl which favour the more placid waters such as this dam, cool behind the hot yellow heads of the everlastings which are a feature of the Park's montane flora.   **170.** Slightly smaller than its European counterpart, the African hedgehog, here with its young, is found throughout the Park, though its nocturnal habits ensure that this insectivorous creature is seldom seen. However, when rains bring forth a mass of millipedes it emerges from the thickets – even in daylight.   **171.** Flap-necked chameleons copulating. The female will excavate a hole in the rain-softened soil of late summer to lay up to 30 eggs which will take almost a year to hatch.

170   171

THE KALAHARI GEMSBOK NATIONAL PARK

HERE IS A THIRSTLAND WITH NO CERTAINTIES, AND WITH FEW EXPLANATIONS.

CREINA BOND. OKAVANGO: SEA OF LAND, LAND OF WATER

As the noose of civilization tightens round untamed nature, so one wildlife spectacle after another has vanished for ever: the massed herds of buffalo on the American Great Plains, the wild horses of the Mongolian desert, the immense schools of whales, even many of the plains animals of East Africa nodding head to tail as they migrated in their thousands. All have been victims of the encroachment of man. Fenced off for their own protection they have become inhabitants of relic islands of wilderness in a sea of habitation. Here the inexorable spiral of management begins, compensating at one point, adjusting at another, in an attempt to reach some equilibrium. From the moment the boundaries of a conservation area are set the process develops a momentum of its own, for the straight unyielding man-made boundaries of the reserved areas almost always cut across the flexible boundaries of ecosystems.

Yet in the Kalahari Gemsbok National Park the grand spectacle of animals moving in their hundreds and thousands still occurs – not regularly and not according to some known scheme. The single most important factor contributing to this phenomenon's persistence into the 1980s is the sheer size of the conserved area. Although the Kalahari Gemsbok National Park is only half the size of Kruger National Park, it flows – unimpeded by fences – into Botswana's even larger Gemsbok National Park. In a splendid act of co-operation across national boundaries the two countries have created a conservation area of some 36 000 square kilometres that ranks as one of the largest in the world.

Many atlases refer to this area as the Kalahari Desert, but strictly speaking it is not a desert at all and ecologists term it semi-arid savannah. The word 'savannah' originated to describe the treeless plains of tropical America but in this southern African context it denotes the sea of grass and scattered trees so typical of the Kalahari dunelands. Kalahari is a European corruption of *kgalagadi*, meaning, 'wilderness', in the tongue of the people who today live in central Botswana. In broader context it is the name given to the almost continuous blanket of sand that extends from the Orange River north through Namibia and Botswana to Angola and Zambia and beyond almost to the equator. It is the largest unbroken mantle of sand on earth, in places 300 metres deep, the reddish grains smothering bedrock from which come strange seismic disturbances that puzzle scientists who record them the world over.

The history of the Park is notably free of

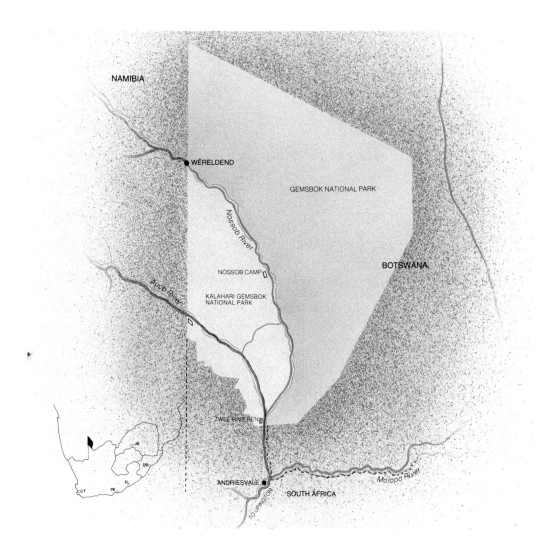

172. *Gemsbok drum a dull tattoo as their hooves pound across the Kalahari's deep sands.*

the political trade-offs and vociferous opposition that distinguishes the early years of the Sabie Game Reserve and, later, the Kruger National Park. The Kalahari has little for man to covet. The sand, coloured a rich Venetian red by an iron oxide layer that persists because there is so little rain which might alter its chemical composition, is deposited by the wind into long, roughly parallel north-west/south-east dunes. They march monotonously across a foundation of calcareous sand or sandstone which breaks the surface here and there, at pans and on the banks of dry riverbeds. Sometimes after a heavy fall of rain these sandstone pans hold water but it soon evaporates under the assault of summer sunshine that pushes temperatures regularly beyond 40 °C and heats the earth to a blistering 70 °C.

Dryness is the distinguishing feature of the Kalahari. Nowhere in the entire Kalahari Gemsbok National Park, nor in its sister park in Botswana, is there any natural surface water. The map defines, with precision, two rivers – the Nossob and Auob – but this gives a false impression for only in years of particularly good rains does the Auob hold a flow of surface water. The Nossob flows on average once every 100 years. Usually the riverbeds may contain the occasional pool – for a few months. . . or

weeks. . . or days. In most years the rivers are entirely dry.

The Nossob defines the boundary between the Park and its neighbour in Botswana, while the Auob to the south joins it a few kilometres before the Park's entrance gate at Twee Rivieren – 'two rivers'. The two rivers (or more accurately the two riverbeds) then join the equally dry Molopo which links with the Kuruman and, in times long past, emptied into the Orange River on its way to the Atlantic Ocean. Today any water that may flow from this tributary drains away beneath dunes that block its path near the Orange. In an earlier age, the wetter periods of the Pleistocene ensured that these rivers merited the name.

The riverbeds and dunes form two distinct habitats. In the riverbeds grow the camelthorn tree, now *Acacia erioloba* but formerly known by the more romantic name *Acacia giraffae*, probably from the preference giraffe show for the foliage and pods which they pluck with their leathery lips, apparently unscathed by the camelthorn's vicious, paired spikes. These trees have come to symbolise life's endurance in the Kalahari: they are hardy, slow-growing and armoured by their thorns. Locals claim that when the camelthorn bears more than its usual harvest of pods it is a warning of drought. In times past the Bushmen gathered the

white droppings beneath weaver bird nests in these trees and added them as a catalyst to their special brew of beer. The soft red-brown heartwood was powdered and used by the women as a cosmetic.

Today there are no Bushmen to make use of the camelthorn, but it is not spurned. Visitors driving along the Park's roads seeking cheetah or lion sometimes stop to appreciate the smaller but no less intriguing characters that create a community about these trees. In the enlarged base of the ferocious thorns, ants often build their nest. That the thorn base is enlarged – and often hollow – raises intriguing questions. Research in Mexico on the relationship between acacias and the ants that nest in the base of the thorns, provided a startling answer: there the ant and tree have been shown to have followed a co-evolutionary path in which the tree provides the ants with ready-made shelter and nourishment, while the ant preys on voracious insects such as moth larvae which would damage the tree. The camelthorn and its resident ant population may well prove a local parallel, and research is under way.

Man or animals taking shelter in the shade of a grey camelthorn or a shepherd's tree may find themselves set upon by tiny bloodsucking tampans, soft-bodied ticks which are present in great numbers in the sand. Their emergence is

173. *Africa's only true fox peers from a tangle of driedoring, typical thorny plant of the Kalahari. The Cape fox is shy and largely nocturnal, resting up in its burrow during the day and emerging in the cool of night to hunt or forage.* **174.** *Trekking to a waterhole ten kilometres distant, blue wildebeest file down the Nossob River. A towering camelthorn sends roots deep down to water, but the river itself flows rarely although pools may form should rain fall.*

triggered by the respiration of man or beast; carbon dioxide from exhaled breath sinks to the ground where the tampans detect it and immediately 'know' that a meal has arrived. In their hordes they then attach themselves to the host, their mouthparts exuding an anaesthetic so that the host is not aware of the bites and does not try to rub the ticks off. Having sucked their fill, they drop back bloated to the sand to breed. The effect of such swarms can be frightening – and unpleasant in that the bites often become ulcerated – and tampans have been known to weaken animals to the point of death.

Older camelthorns have a con-spicuously thick, dark brown bark beneath which shelter any number of lizards that feed, in turn, on the insect inhabitants. The tree rat takes up residence in the fork of the branches and it may well be that its urine – combined with occasional rain water – eats into the wood, weakening it and making it vulnerable both to winds which rip branches from the trunk, and to the burden of sociable weaver nests. These birds construct their tenement-blocks of grass in the camelthorn. Year after year they add to the edifice, hundreds of birds living in pairs in the well-insulated compartments. Other bird species move into vacated portions of the nest, adding to the commotion and vitality.

In the lower reaches of the Auob and Nossob grey camelthorns can be seen more frequently, with sometimes a pair of tiny steenbok peering from the shade.

The sides of the riverbeds are of limestone. Along the upper Auob in particular they are covered with the driedoring. Here the soil is brackish, and the rich Bushman grass that used to cloak the sides, and still does so in the lower Auob and along the Nossob, is sought by grazers such as the wildebeest and springbok. Over the years it appears to have been overgrazed in the upper Auob and the driedoring seems to have filled the barren spaces.

Another shrub abundant in the riverbed areas is the blackthorn. Like the camelthorn, it provides a marvellous microhabitat for creatures as diverse as the Cape penduline tit and a host of rodents. One of these, the tree rat, stacks the base of the plant with grassheads, possibly to hide from the Kalahari's many birds of prey and as a shield from the night-time cold. The blackthorn is fire resistant, but though the gathered grass is a potential fire hazard, in one of those amazing effects and counter-effects between animal and plant, the rodents' scrabbling round the stem creates a natural firebreak.

In the flatter central regions between the Nossob and Auob the coarse Twa or tall Bushman grass flourishes in conspicuous tufts. A less appetising offering is the suurgrass or Kalahari grass, an annual that secretes an acid, sticky substance that causes open sores on contact with animals and man. When actively growing, the suurgrass's defences keep herbivores away, but when it dies off some animals feed on it. More important is its abundant seed that provides a feast for seed-eating rodents, such as Brant's whistling rat and the short-tailed gerbil, for insects and for birds.

The only people who really knew and had adapted to this extreme north-western region of South Africa were the Bushmen, those intuitive conservationists whose ways were sadly misunderstood and undervalued for so long. Early in this century they had largely fled the area of the Park for Botswana where about 1 000 still live in a manner that is but a faint reflection of their former ways. Those who remained either interbred with newcomers and lost their cultural integrity, or were killed off in anger or from diseases the newcomers brought. Initially the area was settled by white farmers hoping to raise cattle on what must certainly have seemed potentially rich cattle country. Later, beaten by the extremes of climate, they abandoned this enterprise and the area was set aside for Coloured people. This subsequent

attempt at settling the far north-western part of South Africa also failed so that when, in July 1931, the Park was proclaimed it roused little opposition. Few saw anything in the achingly dry, dull red sandy tract of not-quite-desert and this has allowed it to survive as a largely pristine wilderness into the 1980s.

The Kalahari's greatest disincentive in the eyes of man is its lack of water, a resource crucial to our existence and one which city dwellers tend to treat with cavalier disregard. But in the Kalahari water cannot be taken for granted. Though totally random, the rain is critical to the survival of animals and plants. The average for the Park is 230 millimetres a year, but in real terms this means little. The range over the past two decades has been from a negligible 56 millimetres to a maximum of 750. Even in a 'normal' year such as 1969 readings from three different stations in the Park recorded close on average for one, two-thirds at another and less than half at the third – cold consolation for a creature that remains in one place and waits for the heavens to bestow a shower of rain. And so the larger herbivorous mammals of the Kalahari Gemsbok National Park are nomads. They follow the water, they trek to the rain, they move in twenties, in hundreds and sometimes in their thousands in search of a green blade of grass or a windfall of seed pods.

At best it is a precarious existence. The environment is harsh. Where the summer air thrums with heat, the winter skies are clear and at night there is rapid heat loss through radiation. Temperatures drop below zero and frost blackens the pale dry grass. Just before the rains, dry winds course over the dunes, scattering seeds and driving dust devils across the pans. At this time of year most of the antelope of the Park are dispersed in the dunes, seeking the last crisped grass, an overlooked seed or fruit, a mouthful of coarse and lifeless leaves.

On average about 10 rainstorms damp the southern Kalahari each year. Where they fall they transform the landscape, for the plants must react quickly to the favourable conditions if they are to complete their cycle of germination, growth and seed formation. Within days the riverbeds are washed with green. If the rains continue, the flush erupts into a mass of greenery splashed with the yellow, purple and white of flowers. With these rains, first the springbok, then the red hartebeest and blue wildebeest and lastly the gemsbok move in, creating a splendid and rewarding sight for visitors from February to May.

175. *Triggered by a thunderstorm, duweltjies bloom across the dunes.*

None of the seven species of antelope in the Park can afford to be too fussy in their diet, selecting vegetation that not only provides their food requirements but also to a large extent fulfils their needs of moisture.

During the rains the fresh green leaves of grasses and herbs are highly nutritious, but as they dry out their food value drops dramatically. When green, the grasses along the riverbeds are superior to those in the dunes and are richer in important minerals. In the dry season, the leaves of shrubs, bushes and trees as well as seed pods are far more nutritious than grass. And so the obvious solution for the antelope of the Kalahari is to graze in the wet season and browse in the dry. But in life, nothing is so simple. Browse material is widely scattered and often occurs in small discrete clumps making it difficult for a large antelope to find enough sustenance. Even when it discovers a supply, its large mouth makes it difficult for it to select many of the smaller morsels so typical of browse material.

On the other hand there are several distinct advantages in being large in an environment as hot and dry as the Kalahari. It is generally true that amongst related animals, the larger species have a lower metabolic rate than smaller ones and therefore on a unit/mass basis require less energy. Their large mouths may be clumsy at selecting the best tidbits, but they can take in large quantities of abundant, albeit poor, dry dune grass if needs be. An equally important advantage to great size rests in the low surface to

volume ratio. Small animals have a relatively large area of skin to body volume and therefore lose relatively more heat, an important survival factor in the Kalahari's often bitter nights, especially for animals that do not shelter in burrows. Finally, being large allows a creature such as an eland to cover greater distances in search of food.

Of the seven most common antelope in the Kalahari Gemsbok National Park, the eland, gemsbok, blue wildebeest and red hartebeest are large, but only one, the mighty eland, is predominantly a browser. At times there are thousands of eland in the Park, at others hardly any at all, their movements governed by the availability of food.

Food in the form of seed pods is extremely nutritious and the pod-bearing plant ensures this, for the pod must attract – and reward – herbivores to disperse the seeds. An animal such as an eland will be drawn to a pod-laden plant to feed and then will move away, depositing the seeds later with its droppings some distance from the parent plant. There, in the nutrient-rich environment the dropping provides, it may germinate.

Like the eland, the gemsbok makes the most of what it can find, eating both browse and grass, although it is predominantly a grazer. These beautiful antelope with their distinctive facial markings and sabre-like horns do not wander as widely as the eland, but prefer to frequent the dunes in the dry season, coming down to enjoy the riches of the watercourse when rains fall. From time to time they visit the riverbed for another purpose: to find minerals either through eating the limestone soil or from drinking the highly mineralised waters at some of the windmills.

Both the wildebeest with its doleful shaggy head and skittish hindquarters and the hartebeest with its oddly-shaped horns are grazers. They are thus drawn to the richer growth in the riverbeds each summer and, as it dries out, retire to the dunes. Nowadays, however, a number of wildebeest have chosen to remain close to the Nossob and Auob throughout the year even when they are dry and the quality of the grazing is poor. There they pose an interesting management problem and, being the Kalahari, it is perhaps not surprising that the problem should centre on water. When the Park was proclaimed it inherited a number of boreholes drilled by farmers in the riverbeds and equipped with wind-pumps and small reservoirs. This water, originally intended for cattle, was now used by the wild animals. They flourished and visitors to the Park found the reservoirs rewarding viewing sites: the Kalahari can be frustrating for people whose appetite for the wilds has been whetted in more heavily populated parks

176. *Wherever camelthorns grow, the marvellous nests of sociable weavers burden the branches. Dozens of these birds – a pair in each cubicle – may inhabit one huge edifice of woven grass. Often pygmy falcons take up residence, apparently tolerated by the weavers because they ward off snakes which take a toll of eggs and nestlings.*

with dramatic concentrations of game. Subsequently more boreholes were sunk, some away from the rivers, out in the dunes.

But in time the expected ecological tangle presented itself. Ecologists are unable as yet to predict the long-term effects of establishing a permanent wildebeest population round the boreholes. It is likely that their presence has caused local overgrazing and the habitat has deteriorated. Not only will the wildebeest suffer; equally at risk is the gemsbok whose traditional source of food in the area may also be lost so that their numbers may well fall.

The Park authorities are, however, monitoring the situation and have closed certain boreholes to assess the effect on the local environment. The Kalahari ecosystems functioned successfully before the boreholes were introduced and artificial watering-points are in fact unnecessary.

Although some springbok are always to be found along the riverbeds, they reach their greatest numbers there during the rains – when they are first to arrive. They are particularly well-suited to cropping very short veld, but are efficient browsers too and survive well in the dry season.

Springbok treks or so-called 'migrations' are part of South Africa's history. The stories of 'uncountable millions' of springbok moving in massed herds are treated with some suspicion by modern biologists, but it is merely a quibbling over scale. There is no question that the phenomenon existed. Indeed, it occurs even now, though the numbers do not approach those of the last century.

In 1959 the springbok populations around Mabua Sehubi Pan, Kang and Kukong in the central Kalahari began to shift and coalesce. Beginning in May of that year thousands of these springbok began to move south-south-west; the movement continued through to July. This area is, of course, sparsely inhabited and so the springbok were not followed or observed in any scientific fashion. It appears, however, that 80 000 to 90 000 merged in the Aminius Reserve about 130 kilometres from Union's End in the northern Cape after July and another concentration was noted east of Twee Rivieren by the Camel Patrol of the (then) Bechuanaland Protectorate authorities based at Tsabong. The total number of springbok must have been astronomical.

This particular trek did not in fact impinge upon the Kalahari Gemsbok National Park, but it serves to show that such mass movements do occur and that the Park is within their potential range.

Two antelope species found in the Park are permanently territorial and thus remain in one area: the tiny steenbok and the only slightly larger duiker. Both live in pairs in restricted territories which provide enough food and shelter for the year and, as both these antelope are independent of free-standing water, they do not need to leave the dunes to visit the boreholes.

Each species of antelope in the Kalahari has evolved its own means of self-preservation in the face of the remarkable array of predators that live there. All the large African predators are represented: the lion, cheetah, leopard, brown hyaena and its larger cousin the spotted hyaena. Then there are the smaller ones: from the caracal to the bat-eared foxes, the black-backed jackal and the beautiful little Cape fox. A great number of raptors patrol the skies, keeping a look-out for birds and small mammals, for insects and carrion. Duiker and steenbok are at risk from air as well as from land. The steenbok relies on its intimate knowledge of its territory – the site of a burrow, a well-placed shadow – and on its own cryptic behaviour when faced with danger. The duiker deals with the problem in similar fashion but is also capable of a surprising turn of speed.

The other antelope tend to gather in herds, finding safety in numbers. Groups of animals have more eyes and ears to detect predators, the individual's chance of being the unlucky victim is reduced, and there is the added possibility that a predator will be confused by the abundance of potential prey and end up catching nothing.

Eland and gemsbok are capable of self-defence, even against lion, and gemsbok wield their horns with fiercesome effect. Hartebeest and wildebeest rely on a speedy escape from danger.

Despite all their defence mechanisms, however, the antelope of the Kalahari ultimately fall victim to predator and scavenger – either through youth or old age, through sickness or some other disablement. Yet, here as elsewhere, it is the prey that dictates the numbers of predators. In the Kalahari with its highly mobile populations of antelope, this effect is even clearer. The big cats, for instance, cannot move far when there are cubs in the pride and so when the herds move away the lion, the cheetah and even the spotted hyaena may face lean times. Significantly, the most common large carnivore in the Kalahari Gemsbok National Park is the one which relies least on active predation for its food. This is the brown hyaena, a solitary and unobtrusive animal seen only by the dedicated

observer and then mostly in the early morning and late afternoon. It scavenges the greater part of its food and kills only five per cent of its intake, supplementing its diet with fruit, insects and birds' eggs.

The Park is a vital refuge for this rare animal which is classed as endangered by the International Union for the Conservation of Nature and Natural Resources and is thus a Red Data Book species. Hunted out by cattle- and sheep-farmers in the past, it remains in a few last refuges. Even in the existing game reserves and national parks its populations may not be large enough to guarantee the long-term future of this unlovely but ecologically fascinating animal.

The Kalahari is a wonderful area in which to observe them. The open terrain makes it possible to follow them at night along the riverbeds and across the dunes. Researcher Gus Mills has followed both brown and spotted hyaena for thousands of kilometres and has observed many fascinating incidents that may change our view of these much-maligned creatures. Despite their skulking manner, and behind that somewhat hangdog expression is a keen intelligence. Mills describes an occasion when a brown hyaena found an ostrich nest with 27 eggs. In the course of the first two nights it managed to eat its way through seven – the equivalent of 170 chicken eggs. Sated but unwilling to give up the rest, the hyaena picked up the remaining eggs one by one and stored them carefully beneath bushes and clumps of grass a kilometre or so from the nest. Its hiding places were random: just as it knew that other egg-

eaters would soon find the nest if the eggs remained there, so it spread its booty in case of discovery. Even if one or two were stolen, the rest would be safe. Mills was there to observe the hyaena when it returned subsequently to dine on its cache. Although it was unable to remember each site, it did remember the general area in which it had stored something and it systematically searched each likely site until it found the hidden food.

Brown hyaenas forage alone, depending on their acute senses of smell and hearing to lead them to food. They roam over territories of up to 480 square kilometres, defended by a small family group, comprising members of both sexes, centred on the den. A scavenger is essentially an opportunist and foraging individually over a wide area, especially one with the Kalahari's limited and scattered resources, provides the best chance of filling the belly. But the brown hyaenas, at the same time, come together to rear the family young – cubs are at least partially dependent on the adults for food for the first 15 months. Each member brings home pieces of meat. The only notable non-participant in the feeding of the young – or in the maintenance of the territory – is the father of the cubs. Mills aptly dubs brown hyaena fathers 'travelling men'. These nomadic males move through the group territories looking for and mating with any females on heat, and then are on their way once more.

Equally vilified is the spotted hyaena which has been relegated in the public's eyes to a lowly role as mop-up support for

the 'great' predators. Few people acknowledge the hyaena's prowess as a hunter – or the fact that a lion will take over a carcass without a moment's hesitation rather than hunt its own prey.

One evening Gus Mills followed an old spotted hyaena from her den. Stopping and sniffing as she went, after about four hours she found a lioness with a still largely untouched gemsbok carcass. A few minutes later the hyaena began to whoop – 19 piercing calls in a row. Back came a reply and within minutes three other members of the hyaena clan arrived to give muscle to her intended take-over bid for the carcass. Before she was joined by the others, the lion had regarded her with indifference. Now, with four hyaenas on hand, the lioness dragged the carcass further under the tree. The hyaenas at this point closed ranks and, with their full cacophony of whoops, howls, laughs and cackles, advanced on the lioness which began growling in turn. Unimpressed, the hyaenas continued their advance and the lioness, after a vain show of strength, fled at the last moment, outnumbered and unnerved.

The Kalahari spotted hyaena is not always lucky enough to steal a meal; more often it must hunt, pulling down animals as large as gemsbok and wildebeest bulls. But its main prey is gemsbok calves which are born throughout the year and thus offer a continual source of supply. The gemsboks' constant moving about helps them avoid predators, for a single spotted hyaena clan may have a territory of 1 500 square kilometres – a vast area for no more than a dozen adults to patrol and to pin-point a particular gemsbok herd

177. (previous page) *Most numerous of the Kalahari's antelope, blue wildebeest trek continually in their quest for fresh grazing.* **178.** *Though the dun of its body blends with its surroundings, the sheer inquisitiveness of the ubiquitous ground squirrel draws attention. Rearing on its hind legs it scans the surrounding veld and, during the very worst of the day's heat, may be seen with its tail curled forward as a natural parasol.* **179.** *Like most of the Park's antelope, red hartebeest numbers ebb and flow as, without regard for national boundaries, they move in response to changes in grazing, seeking nourishment in areas where rain has blessed the veld.*

with young. As it is, an adult gemsbok is not to be taken lightly and the spotteds are chary of those slashing horns.

Most visitors to the Park watch out for the Kalahari lion, expecting to find an animal different and decidedly larger than others elsewhere. They are likely to be disappointed, for the lions of the Kalahari are no different in appearance to those of the Kruger Park. They are forced, however, to feed on somewhat humbler morsels, especially in the dry season when even the porcupine provides a meal.

In the Kalahari a pride of lions will have a territory in excess of a thousand square kilometres. The nucleus of the pride is usually four to seven related adult females plus their offspring. At about two-and-a-half years of age all the males, and even some of the females, are evicted from the pride and become nomadic, the males usually sticking together in their bachelorhood. But at the age of four they begin to eye the males in possession of a pride and challenge these dominant lions. This is often a violent affair and not infrequently leads to the death of one of them – the contender or the incumbent. Nor does the death stop there. No sooner has the new male settled down with his newly-won harem than he may proceed to kill any cubs in the pride. This seemingly abhorrent practice – at least in human terms – stems from a singularly male imperative to sire one's own offspring and waste no energy in giving shelter to young bearing the genetic imprint of another. Sure enough, the lionesses are soon in oestrus again and just over 100 days later the new litters are born.

The Auob riverbed must be one of the best places in the world to observe cheetah on the hunt. The narrow course tends to concentrate the springbok, the cheetah's main prey in the Park, and once a family group of cheetah has been spotted, a little luck and much patience will invariably bring the reward of at least a hunting attempt.

Another splendid sight is the flocks of sandgrouse and doves which arrive in droves in the early morning and late afternoon at the artificial waterholes, rapidly slake their thirst and depart. By coming so regularly to one drinking-place they are vulnerable to waiting predators, but this is somewhat offset by their massed arrival which presents such a roiling mass that any hungry predator would have difficulty fixing on a likely candidate for dinner – and would be unable to catch more than one or two in the short time the birds are at water.

It is interesting to consider the probability that without the artificial watering-places, both doves and sandgrouse would have been unable to take up permanent residence in the Park, even though there is an abundant supply of seed. They must drink daily. Furthermore, sandgrouse need to take water to their chicks. The way in which they do so reveals intriguing adaptations and solutions to problems. Sandgrouse males carry water by soaking it up into their specially adapted breast feathers. During the nesting period, the males can be seen bathing to fill their feathers and afterwards have been known to carry this precious load to nestlings tens of kilometres away.

180

The presence of frogs is another of the Kalahari's many surprises. In this arid environment, several species survive by means of fascinating adaptations which protect them during the dry season. But if rain falls they emerge, usually from mud cocoons in the floor of the pans, and set about breeding in the ephemeral waters.

For the brief periods when the pans brim with water and the riverbeds are studded with pools, the Kalahari takes an entirely different guise. From far away the birds arrive, winging and dipping across the clear blue surface. Waders stalk the shallows, seeking out the frogs and insects that as if by a miracle appear. Algae lace the water's edges, gradually taking over and in doing so the water loses its pristine shimmer and begins to turn a rich dull green. And as the soup thickens, so the birds continue to arrive. The antelope gather to drink their fill and frolic, the exigencies of water shortage temporarily allayed. And the predators, the lion, hyaena, wild dog and cheetah come too, drawn by the festivity of the pans, by the promise of a meal selected from the masses of animals arriving to drink their fill. Above, the sun stares hot and thirsty.

Gradually the waters begin to recede, and as they do so the concentration of nutrients increases, fuelled by animal droppings, encouraged by the sunlight until the waters turn to slime and sludge. The birds enjoy the fecundity and then withdraw and the pan enters its final stages of evaporation; the clay begins to crack and curl and the animals wander back among the dunes to again disperse.

And it is at times like this that the Park experiences the influx that makes the Kalahari Gemsbok National Park stage for those great African spectacles of thousands of antelope massed together. In the winter of 1979 the resident wildebeest population of a thousand swelled to 90 000 when hordes of these animals gathered from Botswana.

This year, next year, perhaps the year after, conditions may favour such an occurrence again. And those who propose a fence along the Nossob should pause to consider the damage it would do in ways we have not yet even begun to comprehend. But of one thing we can be certain: such a man-made barrier will doom the great springbok trek and the grand wildebeest 'migrations' for ever.

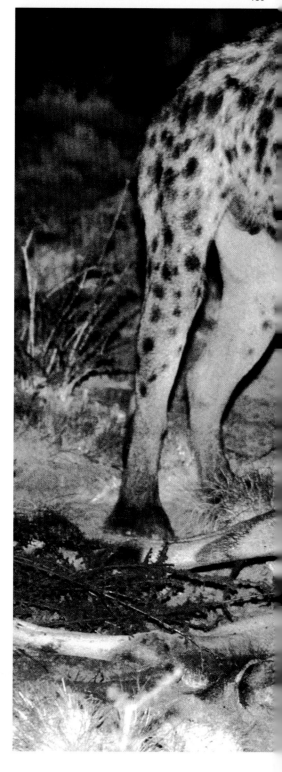

180. *Young ostriches dust-bathing in the talcum-fine sand of the Nossob riverbed. The dust helps rid their bodies of external parasites.* **181.** *Though seldom seen by visitors because it ventures from its burrow only after sunset, the brown hyaena is the arch-scavenger of the Kalahari and covers immense distances each night in search of anything from carrion to tsamma melons. This one found an ostrich nest containing 27 eggs. After gorging on seven, it removed the rest one by one and hid them in the surrounding vegetation for later consumption.* **182.** *Often described as a scavenger only, the spotted hyaena is now known to be a superb nocturnal hunter as well. Working in packs, they run their victims into the ground and then settle down, as this pair has, to an eager meal that begins with the watery content of the gut.*

181

182

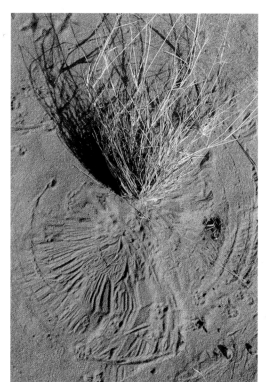

183   184

185

186

183. *Swept by the dry gusting September winds and scorched by years of drought, a tussock of dead grass writes its own obituary in the sand.* 184. *Hunting for termites, this legless skink 'swims' through the red Kalahari sand.* 185. *Incorrectly described as a desert, the Kalahari is more accurately semi-arid savannah revealed here as sinuous dunes cloaked and stabilized by grasses, shrubs and small trees. Another remnant of ancient wetter times is the Auob River, one of many such 'fossil rivers' crossing the Kalahari, and in the Park providing a highly convenient roadway.* 186. *As long as a man's arm and a relative of badgers, skunks and otters, the ratel has a well-earned reputation as a pugnacious fighter when cornered. In the Park these restless creatures are often seen traversing their large territories, digging for insects, and hunting small prey such as this sand lizard – dug up and quickly devoured.* 187. *Ears wide to catch the slightest sound, a steenbok stands alert amid autumn grasses.*

187

188

189

188, 189. *Keenly watched by a white-backed vulture awaiting his chance at the feast below, two black-backed jackals engage in threat displays as they determine who will have first bite. The one on the right, its ears flattened* *and tail down in submission, had taken over the remains of this wildebeest killed some hours earlier by lion. Larger and more aggressive, the jackal on the left arrived to oust the first comer.* **190.** *Returning to take the* *previous night's kill, a lioness drags this gape-eyed wildebeest carcass away from the vultures and jackals into the shade where the cubs are waiting to nibble the remains already savoured by the scavengers.*

191

191. *At full tilt, using its tail as rudder and brake, a cheetah closes on a springbok, chasing it across the broad expanse of the Nossob riverbed. So skilful had been its stalk that it had gone unnoticed until the moment it unleashed its attack, toppling the antelope with a swipe at the hind legs and killing it with a bite to the throat.* **192.** *Although exhausted by its sprint, the cheetah almost immediately began dragging the dead springbok to the shelter of some bushes.* **193.** *Panting in the shade of a driedoring, it kept a look-out for lion or hyaena which often rob cheetah of their hard-won meals.* **194.** *Leaping in alarm, the rest of the springbok herd withdrew. Visitors waiting patiently alongside the Park's dry riverbeds are often rewarded with the adrenalin-charge of a cheetah chase and kill. Slightly more than two decades ago an estimated 90 000 springbok gathered just beyond Twee Rivieren in today's Kalahari Gemsbok National Park. Such astonishing numbers have not been recorded since, but it is possible that this year, next year or 20 years hence they may join up again in their thousands – drawn by some little understood instinct. This Park, flowing as it does unimpeded by fences into the even larger neighbouring Park in Botswana, may yet experience that wildlife spectacle, a springbok 'trek'.* **195.** *Tucking into the meaty, protein-rich haunch the cheetah gulps its meal, wary of the arriving host of scavengers, big and small, which are keen to demand a share.*

192

193

194

195

196. (overleaf) *Fresh meat safely stored in its airy larder, this leopard will return to the springbok carcass over several nights to feed free of interference from scavengers.*

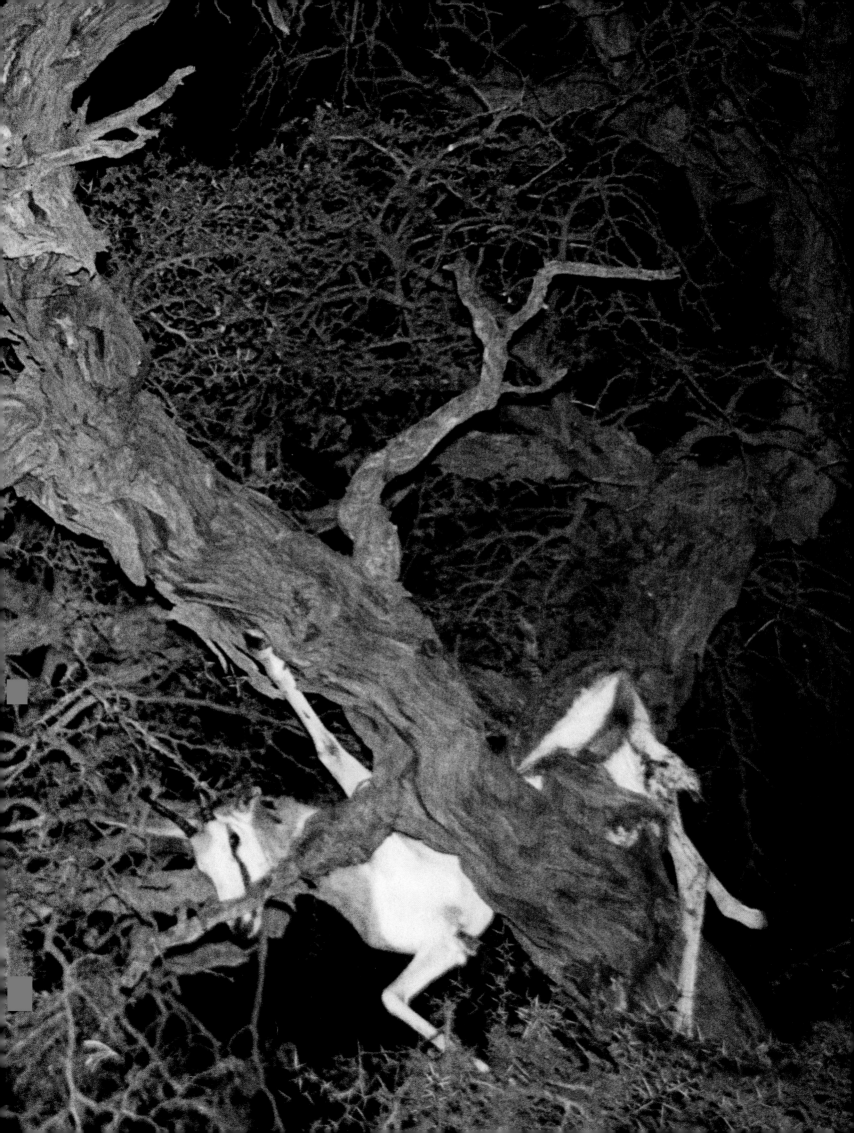

197. *An agama lizard cools off on a thorny acacia twig.* **198.** *Moon set at sunrise and the characteristic reddish iron oxide-coated particles of Kalahari sand assume a deep and plangent hue.* **199.** *The tsamma melon, precious water supply of the Bushman as well as many of the Kalahari's wild creatures. In years of rain the sands are strewn with this fruit: in years of drought they are few and far between.* **200.** *Spotted sandgrouse taking their daily drink. Male birds dunk in the water to drench their specially-modified breast feathers and so carry moisture to their nestlings kilometres away.* **201.** *One of summer's magnificent thunderstorms glowers and grumbles over the Kalahari's aching earth.*

197

198

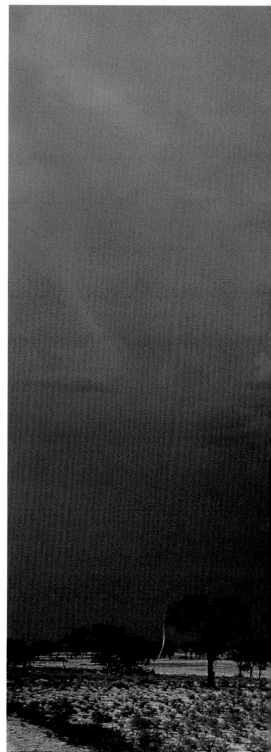

201

199

200

202

203

204

206

202. The wildebeest calving period – usually during the rainy season in December/January – is synchronised over a matter of a few days. The heavily pregnant females come together in calving herds and quite suddenly the young are born, flopping wet and crumpled onto the grass.
203. Only minutes into this world, the calf begins to stand, urged and nudged by its anxious dam. Danger is all about, for this is the predators' great annual feast and hyaena, cheetah, wild dog and lion are close at hand to take their fill.   **204.** The placenta still bulges from this mother as she and her calf nuzzle, imprinting their smells and voice on one another. These few precious, dangerous moments of bonding are vital, for soon they will join the herd and both must recognise the other for the calf to suckle and find safety.   **205.** Fifteen minutes old and barely dry, the calf keeps pace with its mother as she leads it back to the safety of the herd.   **206.** Many eyes watch, ears listen but the danger is past, for the predators have eaten their fill and lie distended and disinterested in the shade. Such is the marvel of nature: some of the calves have been taken but many others may survive.

205

208

209

207. Resting up, a lioness and her cubs take advantage of an acacia's shade. She keeps watch for danger which may well take the form of the Kalahari's nomadic and heavily-maned males. Should one of the nomads oust the pride's dominant male, he immediately sets about killing the cubs of the deposed master. Within days the females are in oestrus again and he mates with them, ensuring that the youngsters he protects will bear his genetic stamp.   **208.** A lioness stalking in sun-bleached Kalahari grass.   **209.** Odd bed-fellows, these porcupines will regularly share burrows with brown and spotted hyaenas which in turn often commandeer them from aardvarks – the original excavators of many such dwellings in their search for termites and other prey – and may lose them to warthog.

210

210. Red hartebeest males, their horns locked, tussle in the Auob riverbed.   **211.** Highly nomadic – as are most Kalahari antelope – eland on the move.   **212.** Most antelope in this harsh and arid environment are able to survive without drinking and draw most of their moisture from their food. Water-loss remains a problem, as does temperature for creatures such as these gemsbok. So successful are they in utilising and maintaining body moisture that urination frequently comprises little more than a few drops. Another adaptation is the fine network of blood vessels in the nose through which blood destined for the brain is cooled, while the body temperature soars during the heat of the day.

211

THE LANGEBAAN NATIONAL PARK

FOR YEARS MAN HAS USED THE SEA AS A SINK AND ASSUMED THAT ITS
VAST SIZE WOULD GRANT IT IMMUNITY
GEORGE BRANCH. THE LIVING SHORES OF SOUTHERN AFRICA

* THIS PARK IS DUE TO BE PROCLAIMED SOON

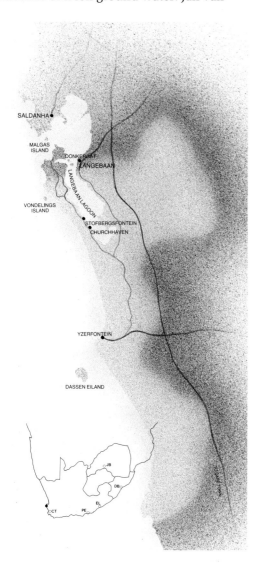

More than 450 years ago Saldanha was named in honour of Antonio de Saldanha who was believed to have taken on water there. This early reference to *Aquada da Saldanha* – the watering-place of Saldanha – clearly bears no relation to reality, for the land round the bay is parched and there is an almost total absence of fresh ground water. Jan van Riebeeck almost 150 years later described the same stretch of land. He wrote: 'There is no land in the whole world so barren and unblessed by the good Lord.' Yet in 1976 the Minister of Environmental Planning and Conservation – not a man prone to flights of fancy – was to describe Saldanha and Langebaan Lagoon as 'the jewel of the west coast'.

Strangely enough, these seemingly opposed opinions are not only reconcilable but symbolize the very features that epitomise this remarkable area: the tranquil haven provided by the calm waters of Saldanha Bay and Langebaan Lagoon, and the harsh aridity of the surrounding land.

It is undoubtedly a unique environment. Professor Roy Siegfried, the Director of the Percy FitzPatrick Institute of African Ornithology, claims 'there is nothing remotely similar anywhere else in southern Africa'. And in recognition of its singular attributes, this jewel has been added to the nation's treasures and proposed as the Langebaan National Park.

To understand why this environment is so unusual, we must pry into its geological history. Both Saldanha Bay and its extension, Langebaan Lagoon, have been created by dramatic changes in sea level over millions of years. Periodic warming of the earth led to the melting of the polar ice-caps, unlocking huge quantities of water which flooded the seas, raising ocean levels up to 140 metres higher than they are today. But during the periods when glacial cloaks shrouded the earth, the ice-caps built up again and the seas receded, often to levels much lower than the present.

These successive advances and retreats of the sea have been central in the creation of Saldanha Bay and Langebaan Lagoon. At times when the sea retreated, massive barrier dunes were built up along the coast. The most recent advance of the sea – about 9 000 years ago – broke through these dunes between the rocky headlands that flank the mouth of modern-day Saldanha Bay. These granite outcrops resisted further erosion so that while the sea flooded through the gap and filled the space behind the dune barrier, thus creating Langebaan Lagoon, the remaining dunes were left intact where they form the backbone of Donkergat Peninsula which now separates the lagoon from the open sea.

From these origins emerges the special character of the lagoon; where almost all other lagoons are created by rivers cutting through to the sea, Langebaan has been formed solely by the rise and fall of the oceans in prehistoric times. Because no river feeds it, Langebaan is not an estuary. This, too, is a key feature, for while the organisms living in estuarine lagoons must cope with the stresses of changing salinity with each rise and fall of the tide, and are subjected to the vagaries of floods and drought, the waters of Langebaan and Saldanha scarcely differ in salt content from the open sea.

Langebaan has a serene beauty which even summer's south-easter winds cannot mar. This same calm extends to the underwater environment where a myriad organisms thrive. More than 550 species of invertebrates occur in the lagoon and the bay – almost double the number in any other lagoon in South Africa.

But calm and constancy are not in themselves sufficient to support such a diversity of animal life. Animals must be

213. *Inquisitive jackass penguins waddle up to the camera on Marcus Island at the entrance to Saldanha Bay. An estimated 10 000 of these birds breed here.*

supplied with food. In the open sea microscopic floating plant-life (phytoplankton) forms the basis of the food web. But measurements show that although phytoplankton is abundant in the rich, upwelled waters of the west coast, and even in Saldanha Bay, its density drops dramatically in the lagoon – as the crystal blue waters so clearly attest.

There are several reasons for the low levels of phytoplankton in the lagoon. In the channels at its mouth lie immense beds of tightly packed white clams and it is probable that they siphon off substantial amounts of phytoplankton as the tide rises to fill the lagoon. In a sense they are a living filter, and the heaps of cast-up clams on the beaches give some idea of their numbers and therefore their effect. But more important than the clams are the rooted plants such as salt marshes, sea-grasses and reedbeds which flourish around the edge of the lagoon and drain nutrients from the water, reducing the levels of nitrates in particular. Robbed of essential nutrients, it is impossible for the phytoplankton to thrive and the salt marshes and other rooted plants assume greater importance, serving in the words of one researcher 'to underpin the entire food web of South Africa's richest lagoon system'. But these salt marshes are also extremely fragile: two years after a beach buggy had ridden across a section of one, its tracks of compacted and almost lifeless mud were still visible.

In spite of the rich growth of plants in the shallows, remarkably few animals feed directly on them. Each year the plants flourish in spring and early summer, only to die back and slowly decay. It is in this death that the plants sustain animal life, for their dead tissues are colonised by teeming swarms of bacteria which break down the plant matter into more easily digested particles. This detrital soup of decaying matter and bacteria provides the main source of food for the worms, snails, prawns, clams and host of other invertebrates that inhabit the lagoon's sandbanks. Some, such as the foot-long bloodworm and the sand- and mud-prawns are well-known; others are less conspicuous but equally important. One species of minute snail, *Assiminia*, occurs in countless thousands near the top of the shore where wading birds such as curlew sandpipers feast on them.

Decaying particles of plant-life tend to be thinly spread on the mud, and so some animals have developed means of concentrating their food. The humble bloodworm is an intriguing example. It excavates a U-shaped burrow through which it circulates water: the water oxygenates the sand and also allows organic particles to be trapped between the grains. Under these conditions, unusually dense bacterial populations build up, only to be eaten in turn by tiny protozoa. This culture of micro-organisms is then consumed by the bloodworm which thus 'farms' its own food supply.

The tiny snail mentioned earlier has even more bizarre feeding habits. Its faeces, rich in carbohydrates, are soon colonised by bacteria. Once this has happened, the snails can feed on their own faeces, taking their food requirements from the bacteria. In this way the faeces can be recycled again and again. Appealing this diet may not be, but nutritious it undoubtedly is.

Oysters were once abundant in the lagoon but, about 9 000 years ago, they became extinct when sea levels changed and silt settled on the lagoon bed. More recently the Fisheries Development Corporation has successfully reared oysters (although a different species to the fossil ones) at Churchhaven on the western bank of the lagoon. Attached to special trays, they thrive there. With the proposal of national park status for the lagoon, however, a debate rages as to whether this should be allowed to continue.

While Langebaan is no longer home to any of the dramatic predators such as lion or crocodile, its multitude of invertebrates supports a host of higher predators. Curiously, there are few fish, but enormous numbers of sandsharks shuffle in the mud where they lie camouflaged, only to shoot out disconcertingly from beneath your feet. The only suggestion that might explain the small numbers of fish is the clarity of the water, for estuarine fish prefer turbid water, presumably because it allows them to conceal themselves from larger predators.

If there are few fish in the lagoon, the birds more than make up for this, for they occur in their thousands upon thousands. François le Vaillant, one of the earliest naturalists to observe the area, remarked on 'the impenetrable cloud of birds of every species and all colours'.

Cormorants, gulls, sandpipers, sanderlings, knots, turnstones, plovers, gannets and flamingos are among the hordes of birds that flock to the lagoon. In summer about 55 000 birds populate Langebaan Lagoon and of this number more than two-thirds are curlew sandpipers. Each year vast numbers of Arctic migrants leave their darkening breeding grounds in Siberia and Greenland and fly south; for many the end of the journey is Langebaan. Perhaps a quarter of all the waders that visit South Africa make for the lagoon and for this reason Langebaan is not only of local interest, but also of international significance as a vital link in the migration route. In Siberia and Greenland enormous tracts of land have been set aside to preserve the areas in which these birds breed. But this move will be in vain unless we are also prepared to protect the southern feeding grounds where the birds spend half the year fattening for the arduous return journey to the Arctic.

Birds must have an immense impact on the lagoon's ecology. Each year they consume some 150 tonnes of tiny creatures, accounting for some 500 million individual animals. About a third of this is returned to the lagoon in the form of guano – 44 tonnes of fertilizer that nurture the salt marshes upon which the birds ultimately depend for their own food.

The multitude of birdlife is attracted to Langebaan not only because of the bounteous feast it offers, but also because of the protection it provides on several islands lying at the entrance to the Bay. Here almost a quarter million seabirds find predator-free roosting and nesting grounds. Schaapen Island, for example, has one of the largest known breeding colonies of black-backed gulls. Marcus Island has the single largest resident population of the rare black oystercatcher in South Africa, while an estimated 10 000 jackass penguins nest there in the densest colony of these birds in the world.

Most spectacular of the bird islands is Malgas, nesting ground for vast numbers of gannets and cormorants. Unkindly and inaccurately described by early Dutch seafarers as *malgas* – 'mad geese' – gannets are actually among the most handsome of seabirds, sleekly coifed in buff and with startling blue eyes, long white necks emphasized by a dark stripe down the throat.

Nesting in the crowded colonies such as the one on Malgas is fraught with conflict, and the gannets defend their nesting sites against intruders, stabbing out with their bills and sometimes inflicting bloody wounds. Because of this, gannets have evolved complex behavioural rituals that allow them to communicate and to recognise their own mates. Something of the importance of these rituals becomes evident when the birds thread their way to the 'runway' – that part of the island left free of nests and from which the birds can get up speed to launch themselves into the air. Reaching this runway is potentially perilous; the birds must waddle between the nests of other birds, running the gauntlet all the way. To signify peaceful intent, they hold their heads vertically upwards – 'sky pointing' to appease antagonists. It is interesting to see how the neck-stripe emphasizes the point, form and function strikingly combined.

When a gannet flies back to its nest after feeding at sea, it pinpoints its mate's call from the seeming cacophony. Then at the nest the two perform an elaborate greeting as part of the recognition ritual, bowing their necks to left and right and arching over the mate.

Together with the cormorants and penguins, the gannets produce a copious carpet of guano. Accumulated over hundreds of years, it made a pungent mantle up to 10 metres thick.

As early as 1845 jetties were erected to allow access to Malgas Island so that this 'white gold' could be harvested. This early plunder undoubtedly affected the birds. The penguins suffered most, for they burrow in the guano to protect their chicks from the sun, and from the attentions of marauding gulls. But perhaps equally damaging was the loss of sanctuary. Not only did humans come onto the island to collect guano, but they also collected thousands of eggs which were regarded as a delicacy.

After the initial frenzied greed when most of the guano was stripped away, the birds slowly recovered and, from 1894 onwards, the guano was harvested more rationally. Care was taken to minimise disturbance of the birds and to ensure that the guano was collected after the breeding season. From that date on a harvest of about 700 tonnes was taken each year from Malgas Island. This annual guano harvest has also fortuitously provided us with a remarkable long-term record of how the population of birds has changed from year to year, for the yield is proportional to the number of producers. There have been fluctuations, but the yield remained remarkably constant from 1895 until the mid-1960s, when it suddenly fell. In 1977, for instance, the yield was down to a mere 152 tonnes. This rapid decline is no doubt a reflection of the commercial over-fishing of pelagic fish, particularly the pilchard, for this fish makes up the birds' main diet – and, indeed, fisheries scientists have used the yield of guano as an index of just how many fish there are.

A more recent threat faces the birds on Marcus Island. In 1976 a causeway was completed between the mainland and the island, serving as a breakwater for the major harbour construction in Saldanha Bay. As a result, Marcus is no longer an island and can no longer protect the birds from terrestrial predators. The danger was recognised and pointed out four years before the causeway was built, yet six years were to pass after its completion before the authorities put up a predator-proof wall. By that time grey mongoose, water mongoose, genets, porcupines, bat-eared foxes, Cape foxes, meerkats, a domestic cat, a rat and fieldmice by the hundred had found their way over the 1,3-kilometre causeway, and there was mayhem among the birds on the island.

The Langebaan National Park will not only safeguard the lagoon but will also protect an area of unique West Coast strandveld vegetation on the mainland and on Donkergat Peninsula, as well as a stretch of coastal fynbos. Parched though this land may be, and brittle and thorny the plants, the annual rains ring a breathtaking transformation. In late winter and early spring the sand dunes are carpeted in flowers: daisies, vygies and many more. And after the vibrant display dies down, the plants speedily set seed before summer's angry heat burdens them once more.

The land supports many small mammals as well: the dainty big-eared steenbok is commonest of the antelope, but smaller numbers of grysbok and duiker can be spotted too. Because of their secretive ways, small predators such as mongoose, genets, caracal and jackals are more rarely seen, but the delightful bat-eared foxes appear frequently on Donkergat Peninsula. Although these feed mainly on insects and are more the farmer's ally than foe, they have been killed off along with the many, many more animals that once must have occurred here but have since been exterminated by hunting and by the changes to habitat agriculture brings.

In the mid-1600s, visitors to Saldanha reported seeing rhinoceros, elephant, eland, lion and numerous antelope. Lord Somerset later used the house *Oostewal*,

214. *Malgas Island teems and roils with gannets, particularly during the breeding season. In the crowded colony a long strip is left vacant as a runway for the birds to gain speed before launching into the air. When a gannet 'sky-points', the dark stripe down its throat emphasises this posture signalling non-aggression as the bird makes its way through the gauntlet of threatening beaks to reach the runway.* 215. *On the seaward side of Langebaan Peninsula spring flowers evoke calm while stormy seas lash the beach.* 216. (overleaf) *'Jewel of the West Coast', the unique ecosystems conserved at Langebaan nurture more than 550 species of invertebrates, the wonderful west coast strandveld vegetation and, on the islands, a myriad birds. A wealth of small creatures – snails, worms, prawns, clams – thrive on decaying detritus released during the life-cycles of plants growing in the immense salt marshes in the foreground. These creatures in turn support tens of thousands of waders, mostly northern migrants that gather here in summer. Some 30 000 curlew sandpipers fly south from Greenland and Siberia, where their breeding area is protected, to Langebaan Lagoon to feed and fatten.*

217. *Kelp (black-backed) gulls on the edge of the Lagoon. The bird on the left holds a snail foraged from the sand as the tide retreats, exposing stretches of sandbanks teeming with life.*   218. *Sarcocornia, a major component of the fragile salt marshes that underpin the entire ecosystem of the Lagoon. Each year they flourish in spring and early summer, only to die back and slowly decay. Their dead tissues are colonised by bacteria that break down the plant matter, providing a detrital soup on which the innumerable invertebrates flourish.*

which still stands on the banks of Langebaan Lagoon, as a hunting lodge.

Spectacular as the 17th-century big game must have been, it is eclipsed by the astonishing creatures that roamed here in prehistoric times. About 15 kilometres north-east of Langebaan exceptionally rich fossil sites have been discovered and intensively studied by palaeontologists from the South African Museum. Their research has revealed an extraordinary procession of prehistoric beasts, together with an enormous amount of information about the area's climate long ago. It daunts the imagination to stand on Saldanha's parched earth today and try to visualise this same landscape cloaked in lush forest sustained by high rainfall; to imagine such extraordinary creatures as the short-necked giraffe-like *Sivatheres*, musk oxen, giant pigs three times the size of bushpigs, sabre-toothed cats and bears that weighed about 750 kilograms – four times the weight of a fully grown lion.

The fossils unearthed at Langebaanweg provide a record of life spanning almost 24 million years. During this time, changes in sea level and in temperature have had a dramatic effect on climate, both here on the west coast and all over

the world, and upon the evolution of life.

The inexorable changes in climate left their mark: freezing of the polar ice-caps shrank and cooled the oceans. The colder seas released less moisture so that rainfall diminished and the once luxuriant forests of the west coast were replaced by open grasslands and, gripped by increasing dryness, the land took on the arid aspect we see there now.

In the process of this change most of the giant mammals became extinct, leaving their fossilised bones which have been discovered by a trick of fate. In prehistoric times the Berg River had a different course to that it takes today, for it originally opened into a shallow bay that lay inland of the modern town of Saldanha. As all rivers do, the ancient Berg carried minerals down to the sea. But, this fresh mineral-rich water is only one in a combination of factors involved in the complicated geochemical processes that account for the immense deposits of tiny phosphate pellets. These originated in the sea where millions of tiny dead marine organisms built up in the presence of nutrient-rich upwelled waters which were warmed in the shallows. As they accumulated they buried the bones of

dead animals and eventually were to form deposits up to 20 metres deep. Limited mining of these rich phosphate beds took place in the 1950s, but it was only when Chemfos Limited expanded their operations in 1965 that the unparalleled extent of the fossil beds came to light – beds that make the area of international significance and were described by Dr Q.B. Hendey of the South African Museum as 'one of the most prolific fossil sites in the world'.

The Saldanha-Langebaan region is one of contrasts. But perhaps the most significant is between the sleepy lagoon, with its wheeling clouds of birds, and the massive industrial developments taking place at Saldanha. Visitors to Langebaan experience this counterpoint: the serene beauty of the lagoon on one hand and the angular constructions of the iron ore terminal clearly visible at Saldanha on the other.

If the two were independent we could afford to ignore this juxtaposition, but they are not. Langebaan Lagoon acts like a lung, partially filling and emptying with each tidal cycle, exchanging water with Saldanha in the process. Consequently whatever happens to the waters of Saldanha has a bearing on the ecology of the lagoon. Already it has been affected.

The finely-ground iron ore is dust-like and, despite rigorous precautions, inevitably forms clouds when ore is being loaded. Iron ore does not dissolve in water, and so it tends to sink close to the

harbour wall and blankets the seabed where its build-up could have serious repercussions. However, its effect has been relatively local to date. And there have been lighter moments in the heavy debate on this question of air and water pollution. Gulls delight in floating in the sheltered pools near the base of the harbour wall, where they soon become a bright brick red as a result of the iron ore particles. Unharmed, but slightly absurd, these exotic-looking seabirds excited so much comment that the harbour authorities draped nets over the pools to exclude the birds. A sign next to one pool bears the peremptory instruction: *Waarskuwing aan alle seemeeus. Die swem in rooi water is streng verbode.* (Warning to all seagulls. Swimming in red water is strictly forbidden.)

Other developments associated with the harbour wall have posed more serious threats. To construct a breakwater between Marcus Island and Saldanha, massive amounts of sand were dredged from the bottom of the bay and were dumped in the channel between island and mainland. Most forms of life in the sediment were obliterated. More damaging were the copious quantities of silt that became suspended in the water when the sediments were churned. Some silt found its way into Langebaan Lagoon where, in the tranquil waters, it settled, blanketing rocks and killing numbers of octopus in the process. In Saldanha Bay the suspended silt cut out light and

limited the supply of oxygen. Among the casualties was the once dense bed of *Gracilaria* seaweed which was silted over and all but eliminated. *Gracilaria* is a major source of agar, widely used in bacteriological and pharmaceutical work as well as in the manufacture of glues and jellies. And the price was paid: in 1973 the seaweed harvest earned R1 000 000. After dredging began in 1974, the seaweed yield was not even 10 per cent of this and the industry collapsed.

To allow giant ore-carriers access, a deep approach-channel had to be blasted through the bedrock of the bay. Explosion after explosion was marked by an eruption of water hurled 100 metres into the air by a blast of 200 kilograms of high explosives. Untold numbers of fish were killed, their bodies floating on the surface after each blast. And then the penguins and cormorants arrived to dine, only to be doomed by the next blast.

In spite of the fact that an Advisory Committee for Ecological Studies had been appointed by the government, their advice was not sought before blasting began. Thousands of birds had perished before the matter came to their attention. The bloodshed had been completely unnecessary: by playing tape-recordings of killer-whale calls and by discharging revolvers, it was relatively easy to scare the birds away before each blast, and to arrange for blasting to take place at a time when the birds were least likely to be feeding. Simple measures such as these

ended the massacre, but it is sad commentary that ecologists were not consulted earlier.

The message is clear, research and scientific advice are vital. In Langebaan Lagoon we have a unique system; one that is biologically rich and at the same time one that is fragile and extremely vulnerable. An important step has been taken towards its conservation, for it is to be proclaimed a national park. But proclamations do not ensure preservation.

We can never forget that Langebaan shares its waters with an industrial giant in the making. Nor should it be imagined that it is only industrial developments that must be controlled to ensure survival of this remarkable area. As human numbers build up, recreational use of the area will in itself be a major problem. Mere overuse and trampling – even by the best-intentioned visitors – can be sufficient to destroy a fragile ecosystem such as this.

219, 220, 223. *Arid and parched for most of the year, the vegetation surrounding the Lagoon unfolds in a spectacular springtime display. Although rainfall is niggardly, in wetter years an exuberance of Bokbaai vygies, daisies, gazanias and other flowers greet visitors.*
221. *Scavenging sacred ibises rise gracefully into the air, revealing the melanin-blackened edges of their wings which are strengthened by the presence of this substance.*

222. *Frequently hunted and trapped, the bat-eared fox is a much-maligned creature, for it feeds mainly on insects and rodents, and is more the farmer's ally than his foe.* **224.** *Donkergat Peninsula seen across the Lagoon's serene waters which owe their crystal clarity in part to the immense and densely-packed beds of white clams that filter the water as it ebbs and flows with each changing tide.*

219 220

221 222

223

225

225. *As valuable as the Lagoon itself are the several islands lying at the entrance to Saldanha Bay. They offer predator-proof roosts and nesting sites for an estimated three-quarters of a million seabirds – including the oystercatcher with its crimson beak and eyes, jackass penguins, cormorants and gannets.* 226. *Squawking and draping her wings protectively over her nest, a bank*

226

227

cormorant on Malgas Island makes her displeasure clear. These large, black seabirds build their seaweed nests on various offshore islands along the cold West Coast. The litter around the nest reveals these birds' predilection for crustaceans – including the South African rock lobster which inhabits kelp beds. **227.** Wall-to-wall Cape gannets at Malgas Island, site of one of the largest breeding colonies of this species. Nesting here in their tens of thousands, these handsome birds with their buff heads and startlingly blue eyes create a spectacle noted by early visitors to this coast. They recognised the wealth of guano and by 1845 jetties had been erected to allow the harvest of 'white gold'. Since an initial frenzy when these riches were ruthlessly exploited, the guano has been collected on a more controlled basis. The annual harvesting of guano has also provided a record of fluctuations in the bird population over the years which, in turn, has been used to gauge the decline in the shoals of pelagic fish along this coast – one more sad tale of man's greedy over-exploitation of a once rich natural resource.

228

229

228. A kelp (black-backed) gull 'intimidates' young Cape cormorants, causing them to regurgitate and thus provide an easy meal. The youngsters are themselves fed by their parents, which fly daily sorties to catch fish for their offspring.   229. Jackass penguins flip-flop to the water's edge on Malgas Island. Their loud night-time braying reverberates from the island and accounts for their common name.   230. Ungainly and lumpish on land, jackass penguins come into their own in water where their sleek torpedo-shaped bodies and stubby wings propel them swiftly and safely even through this rough surf.
231. A scruffy-looking jackass – its moulting plumage stained green where it slithered belly-down over algae-covered rocks.
232. (overleaf) Visitors to Langebaan, lesser flamingos flock pink in flight across its welcoming waters.

230

231

233

234

235

236

PHOTOGRAPHER'S

ACKNOWLEDGEMENTS

At no other time have I relied on the goodwill of so many individuals as in the case of this book.

The assistance and cooperation I received from the National Parks Board at all levels was truly amazing. In particular I would like to thank the Chairman of the Board, Prof. F. Eloff; the Chief Director, Mr A.M. Brynard; the Head of Information and Research, Mr Piet van Wyk; the Head of Southern Parks, Dr 'Robbie' Robinson; and Chief Warden of the Kruger National Park, Dr 'Tol' Pienaar. On many occasions they interrupted busy schedules to answer my questions, offer advice, and to convey their enthusiasm for this project.

The park wardens, rangers, research officers and information officers were wonderfully helpful in ensuring that I saw as much of each park as possible. Many of them worked long outside their normal hours to help, often inviting me into their homes, or to share a campfire circle. Their spontaneous kindness and friendly cooperation I value deeply.

A great many others gave freely of their time, skills and hospitality, and helped me in a great many ways. Space is hopelessly short to allow me to list them all, or the nature of their assistance, but some of these kind people are: Lindes Basson, George and Margo Branch, Bill and Tessa Branch, Peter and Jane Betts, Alan Davidson, Pat Evans, Karen Freimond, Russel and Bonnie Friedman, Richard Goss, John and Wendy Greig, Peter and Claire Johnson, Edouard le Roux, John and Amy Ledger, Paul Martin, Carl Meek, Tony McEwan, Harold and Tiny Mockford, Ken Newman, Patrick and Marina Niven, Rod and Colleen Patterson, Tony and Bea Petter-Bowyer, Mark Read, Moppet Reed, Christine Renwick, Michael Rosenberg, John Skinner, Philip and Althea Steyn, Rudi van Aarde, Adolf and Liz Waidelich, and the staff of the Percy FitzPatrick Institute.

There are several other people to whom I owe a particular debt of gratitude: to the production team at Struik, for their enthusiasm, energy and patience – particularly to René, Pieter, Wim, Walther and Bunny; and to Lesley Hay and Lyn Wood for keeping my office running smoothly during my many periods of absence; and to members of my family who by design or coercion took part in one way or another, especially Joan Lawrenson, Maudanne Bannister, and my children Andrew and Sue for their helpfulness and humour on those occasions they could accompany me into the field.

And finally to my wife Barbara for staying home for David and Patrick, and for her love and encouragement.

ANTHONY BANNISTER, JOHANNESBURG 1983

AUTHOR'S

ACKNOWLEDGEMENTS

Only one hand moves the pen, but many minds guide the writer. In writing this book I have been privileged to have been joined by many fine minds and, more importantly, many fine people. First among them is Professor Roy Siegfried to whom I owe so much. I thank him for his gift of knowledge, his critical and perceptive assessment of the manuscript and for his support. I am also grateful to Peter Schirmer who ironed out the worst of the stylistic wrinkles and added his special flair for the written word; to Tony Bannister with whom I have worked on several books, each serving to confirm the pleasure of working with a photographer of immense talent and a human being of great integrity; to the consultants without whose generous sharing of research the book would have been without substance and authority; to Professor George Branch who brings to his subject such marvellous clarity; and to Walther Votteler whose gifts as a designer are evident on every page and who ranks with the very best in his field of creative book design.

I am grateful to Dr P. van der Walt and Mr P. van Wyk of the Parks Board, and to Dr Anthony Hall-Martin at Skukuza, all of whom read the manuscript in part or in full and made valuable comment; to Dr V.B. Whitehead of the South African Museum for his intriguing information on the insect life of the Kalahari; and to Richard Brooke and Richard Knight of the Percy FitzPatrick Institute of African Ornithology who checked the scientific and common names.

Lastly I must thank my husband for his support, and my children who bore my distractedness with a shrug and a smile.

RENÉ GORDON, CAPE TOWN 1983

233. *A cushion star nestles among blades of sea-grass.* 234. *Cast off during the act of moulting, the carapace of a swimming crab, Ovalipes, lies stranded on the beach. Swimming crabs are voracious predators, feeding on white mussels, plough snails and even small fish.* 235. *Protected from exploitation in a sanctuary, enormous numbers of rock lobsters frequent the shores of Saldanha and its islands, where they feed mainly on mussels. In 1972 more than 10 000 rock lobsters perished as a result of pollution from fish factories at Saldanha, but since then a new system of offloading fish has been instituted and pollution greatly reduced.* 236. *The common shore crab, Cyclograpsus, usually shelters under rocks high on the shore, but in Langebaan Lagoon it scuttles among the salt marsh plants where it is preyed on by otters.*

INDEX

A NOTE ON THE PHOTOGRAPHY

It may be of interest to other photographers to know how the photographs for this book were taken. Most were done over an 18-month period, travelling on normal tourist roads in a small passenger car. Two basic camera systems were used:

A Nikon system (35 mm) comprising two Nikon F3 camera bodies and the following Nikon lenses: 15 mm, 24 mm, 55 mm macro, 105 mm, 200 mm macro, and 400 mm.

A Hasselblad system (6 x 6 cm) comprising one Hasselblad 500 ELM body, and the following lenses: 40 mm, 80 mm, 150 mm and 350 mm.

NATIONAL PARK INFORMATION

For further information and reservations write to: The Chief Director, National Parks Board, P.O. Box 787, Pretoria, 0001
Telegraphic Address: NATPARK, Pretoria
Telex: 3-642 SA
Telephone: (012) 44-1191, 44-1100, 44-1102.

Copyright with kind permission of:

Oxford University Press, Oxford, for the extract from F. Fraser Darling's *Wildlife in an African Territory* (1960);

William Collins Sons and Co. Ltd, London, for the extract from J. Stevenson-Hamilton's *South African Eden* (1937);

R.N. Currey for the lines from his poem 'Remembering Snow';

I.S.C. Parker for the extract from his paper 'Conservation, Realism and the Future', delivered at the Symposium on the Management of Large Mammals in African Conservation Areas, CSIR Conference Centre, April 1982;

Perseus Adams for the lines from his poem 'The Woman and the Aloe';

G. Harrap Ltd., London, for the extract from P.J. Pretorius' *Jungle Man* (1947);

Francisco Campbell Custodio and Ad. Donker (Pty) Ltd. for the lines from Roy Campbell's poem 'The Zebras' from *Adamastor*;

Uys Krige for the extract from his translation of Eugene Marais' 'The Dance of the Rain'.

First published in the UK in 1992 by
New Holland (Publishers) Ltd
37 Connaught Street, London W2 2AZ

Copyright © in text René Gordon
Copyright © in photographs Anthony Bannister, except for the following which are reproduced by kind permission of Peter Betts (27), Clem Haagner (54), Gus Mills (180, 181), National Parks Board (50–53) and Rex Symonds (162)
Copyright © 1992 New Holland (Publishers) Ltd

ISBN 1 85368 185 7

Designer: Walther Votteler
Illustrator: Tony Ribton
Phototypeset by McManus Bros (Pty) Ltd
Originated by Hirt & Carter (Pty) Ltd
Printed and bound in Singapore by Tien Wah Press (Pte) Ltd